# Along Came a Llama
# RUTH JANETTE RUCK

# faber

First published in 1978
by Faber & Faber Limited
Bloomsbury House
74–77 Great Russell Street,
London WC1B 3DA

This edition first published in 2020

Typeset by Faber & Faber Limited
Printed and bound by CPI Group (UK) Ltd, Croydon, CR0 4YY

Illustrations by Paul Work

A CIP record for this book
is available from the British Library

ISBN 978–0–57136–319–3

MIX
Paper from
responsible sources
FSC
www.fsc.org
FSC® C020471

2 4 6 8 10 9 7 5 3 1

# Along Came a Llama

To my dear sister-in-law
Jean Mary Horton

# The Llama

The Llama is a woolly sort of fleecy hairy goat,
With an indolent expression and an undulating throat
Like an unsuccessful literary man.

HILAIRE BELLOC

# Contents

Acknowledgements                                    *page* xi
Foreword by John Lewis-Stempel                           xiii

  1. Away to the Hills                                     1
  2. Zoo Visit                                             9
  3. The First Step                                       19
  4. The South American Beauty                            32
  5. Home Again                                            45
  6. An Anxious Time                                       54
  7. Settling In                                           64
  8. Waiting for the Snags                                77
  9. Mooey                                                89
 10. Llama at Home                                       112
 11. Coat and Slippers                                   124
 12. Llama Drama                                         136
 13. First Birthday                                      148
 14. Bracken and Brandy                                  158
 15. Beginning the Second Year                           172
 16. Some Solid Facts                                    184
 17. Mr Widdle                                           192
 18. Llama Revealed                                      205
 19. A Hot Summer                                        225
 20. The Circus and Some Science                         245

21. The Third Year                        261
22. To Chester Zoo                        271

# Acknowledgements

The practical kindness of others has helped me considerably with this book. In particular I would like to express my sincere thanks to Miss B. Fargher and Mr N. R. Ingman, our vets, and to Dr Antice Evans for scientific facts. For illustrations, my gratitude goes once more to Brian Nicholson for preparing the photographs, not only his own but also those kindly given by John Greaves. And to Paul go my thanks again for drawing the pictorial map and for the sketches.

Hilaire Belloc's poem *The Llama*, taken from his *Complete Verse*, is reproduced by kind permission of Gerald Duckworth & Co Ltd, and of Alfred Knopf Inc in the USA.

Finally, I acknowledge with gratitude the award of a grant from the Welsh Arts Council for the purpose of writing this book and I thank my friend, Mollie Keen, for her stroke of genius in suggesting I apply for the grant.

# Foreword by John Lewis-Stempel

Like its prequels *Place of Stones* and *Hill Farm Story*, Ruth Ruck's *Along Came a Llama* belongs to the 'Back to the Land' vogue in Britain in the 1960s and 1970s, which found its most famous expressions in the book *Self-Sufficiency* by John and Sally Seymour and the television programme *The Good Life*.

Ruck's books depicting life on the 83-acre farm of Carneddi in pluvial, precipitous Snowdonia were simultaneously more authentic and more charming than rival works.

The Ruck story really begins in 1945 when her demobbed father decided to up sticks from suburban Nottingham and relocate to the threadbare grass and hard slate of North Wales. Ruth was seventeen.

Going back to nature, of course, has always been the British way in a crisis; at the end of the Great War there was even government legislation to fund ex-soldiers on smallholdings. The original classic of self-sufficiency was William Cobbett's *Cottage Economy*, written in impoverished, depressed 1821. Cobbett's book itself was indebted in its philosophy to the Diggers of the world-shattering English Revolution.

So, naturally, we are nature-minded again now: post-crash, post-coronavirus.

Luckily Ruck's family took to farming in the mountains of North Wales like a sheep to grass. By the long hot summer of 1976 – the year in which the llama of the title came into their lives – Ruth had married climber Paul Work and taken over the reins at Carneddi. Around her, even in remote North Wales, traditional farming was disappearing. The Rucks were nearly alone in turning hay by hand, in willing individual sheep – known by their faces, their names – to live when ill. Everyone else had been mechanized, chemicalized, industrialized; or sold up. The Age of the Big Farmer had arrived. Ruth Ruck, however, was 'so indoctrinated by the old ways' that Carneddi continued as before.

There is food for thought in these pages (which, by the way, are deceptively literary underneath Ruck's easy, 'chat over the kitchen table' style). Disciples of Big Farming say that these old ways of agriculture are uneconomical . . . which is an odd criticism from an industry only kept afloat by billions of pounds in subsidies, in which farmers lie awake at night counting not sheep but repayments on a tractor that can easily top £100,000.

Of course, I *would* agree with Ruck. I still turn hay by hand with a rake. My tractor is 1956 vintage. (The Rucks' Land Rover, persistently 'up to its tricks', also strikes a chord.) I too talk to our animals.

Ruth admits that the hard work 'put lines on my face and grey in my hair'. She does not romanticize her life, but she is alive to its benefits, such as free-range kids who can swim in the stream, and surroundings which are 'a constant inspiration' in their beauty. The phrase she uses about her style of farming is 'soul-rewarding'.

Is there any soul at all in modern agriculture? Any esteem *at all* for livestock on slats in factory units?

Something else pleasantly, positively, old-style about *Along Came a Llama* is Ruth's mental toughness. In these pages her father dies, her sister Mary dies, and she herself contracts multiple sclerosis. This could so easily have been a misery memoir.

While she does find the MS 'a sentence on me which I found very hard to bear', instead of wallowing in self-pity she grits her teeth, pulls up her socks and puts on her wellies, 'as one always does when farming'. Her husband, Paul – reflecting that he has injected hundreds of sheep, 'so why not his wife?' – gives Ruth her quotidian medicine.

There is a llama in the room. Literally. While the Rucks carry on raising sheep, poultry and Welsh black cattle, they do not quite carry on regardless. They buy a llama from Knaresborough Zoo to 'cheer us up', taking it home in a trailer with no fuss, no paperwork. (You could do fun things like that in the 1970s; today you would need online movement forms and licences in triplicate.)

So, enter Ñusta (Quechan for 'Princess'): a toddler llama, with big eyes and bigger appetite – sometimes for rather unusual fodder. (Tissues. Maltesers. Cherry brandy. Tic Tacs. The *Radio Times*.) Indeed, the very first line is a taster of what is to come: '"Oh, heck," cried Paul. "It's at the sugar again!"'

The Rucks know nothing about South American llamas from the high country of the Andes; but then again, nor do many other people in the Britain of their day. But the Rucks do understand animals, and they do care: so after some hiccups in llama-raising, Ñusta settles down happily at Carneddi, taking pride of place on the hearth rug next to her toy box, 'calm, dignified and beautiful', and making her slight musical moos.

To the Rucks' existence, already full of soul, Ñusta adds character. Even on paper, travelling across the decades, Ñusta's personality entrances. One departs this book a convinced llama-lover, despite the species' camelid reputation for spitting. (Ñusta is far too refined for such behaviour. Usually.)

No dumb animal, Ñusta. Looking into Ñusta's eyes Ruth has the 'uncanny feeling that there was someone in there, trying to communicate'. When Ñusta looked back, she would have seen respect in the eyes of her 'llama mama'.

Care for the environment. Connection with nature. Self-sufficiency. Regard for animals . . .

Certainly, *Along Came a Llama* has a Durrellian recollection-of-family-life-with-unusual-pet spirit about it. But it is so much more. It is a guide to the future.

To a good life.

ÑUSTA'S
WAY
UP TO
CARNEDDI
FROM
CAPEL ANWES
BARN

BRYN
UM

GREEN
PATH

R.W.

# 1 Away to the Hills

'Oh heck!' cried Paul. 'It's at the sugar again!'

We all made a dive at the animal. Her nose was firmly clamped into the silver sugar-bowl. The sugar was disappearing out of the bowl at the rate one would have expected if the nozzle of the vacuum-cleaner had been applied to it. Paul grabbed the bowl and passed it over the llama's back to me. I hastily put it in the food-safe and shut the door. The llama put her ears back, goggled her eyes and pulled ugly faces, pretending that she was going to spit. Then she turned to the sugar that had been spilt and began to vacuum up the last grains with her quick upper lips. Her ears went forward again.

There wasn't much space in our small kitchen with three of us and the llama. The animal was so big now that her body reached from side to side of the room. Her neck was so long that she could touch all but the upper shelves, where the sugar, flour, porridge oats and cornflakes had now taken refuge.

When the llama had finished all the sugar, she took a last look round to make sure that there was nothing else that she fancied and, giving a quiet toot as though to say thank you, she trundled out of the kitchen and into the sitting-room. There was silence for a moment, then a rustling sound, then silence again followed by a muffled thud. I knew, without being there to see, that the llama had gone into the next room and turned over the contents of her toy-box to see if there were any fresh magazines or newspapers in it. Next, not feeling in the mood to eat paper, she had stepped on to the rag rug, positioned her little ballet-dancer's back feet together, knelt first on her right knee, then on the left, sagging a little at the hocks meanwhile, and had finally gone right down with a small thud. Now she would be sitting there in all her glory, with her white neck upright and her draperies of flowing mauvey-grey, fawny-gold, gingery wool hanging down to the mat all round her. There would be no sign of her legs and we called this the 'tea-cosy position'. Then there was a rattling sound. I knew that she was shaking her head rapidly from side to side, causing the long ears to fly round in circles—'helicoptering' we called it. Then there was silence again.

When the dishes were washed, I went into the sitting-room to see her. There is something very restful about having a llama seated on the hearth rug. The animal is calm, dignified and beautiful. Your eyes are attracted towards it again and again, to the long ears, to the small head and the enormous eyes under a fringe of black lashes, to the curve of its body mantled in wool. Our family found that having a llama was a pleasure far greater than we had originally expected it to be.

The llama's presence sprang from the germ of an idea which I had had long ago. As a child, I was deeply interested in animals. We lived in the city of Nottingham but I would spend hours gazing at the cows, farm horses and sheep whenever we went into the country. In time I had a dog of my own, a rabbit, six hens and two goats in our large garden in the city. I enjoyed this mini-farm but it wasn't quite enough. Somewhere there is a notebook with a list in childish handwriting.

> *I want a pony*
> *I want a cow*
> *I want a sheep*
> *I want an elephant*
> *I want a llama*

The elephant has never become a reality, but the others have. It is strange how dreams often come true in the end.

The years passed and the child's urban play farm was transposed to North Wales and became the real thing. A number of arbitrary events caused this. The War ended on my seventeenth birthday. My father, who had been in the National Fire Service, was made redundant. I had a sudden and serious illness which nearly cost my life. Afterwards my parents took me for a holiday in North Wales to recover. There we heard of a small mountain farm which was for sale cheaply. My father was a man who didn't fit into the mould of suburban bread-winner which circumstances had imposed on him till then. My mother had an adventurous and unquenchable spirit and Fred, our ex-nanny, family retainer and second mother to my sister and me, was a countrywoman. I longed to farm.

The wild, intoxicating idea of buying the little farm, casting aside our previous life and becoming hill farmers suddenly seemed alluring, suddenly seemed possible and suddenly became reality. We sold our Nottingham house and moved, lock, stock and barrel, goats, hens, bees and all to the mountain farm of Carneddi, on the foothills of Snowdon, looking down to the sea seven miles away. It was December 1945.

It was, of course, a mad thing to do. Everyone said so. We had little money and no experience, but it was a colossal adventure into which we all entered with enthusiasm. It was in the days before people used the words 'environment', 'ecology' or 'conservation' so frequently. The idea of subsistence farming was so unfashionable that it seemed quite irrelevant. All that was still twenty years away. But, in 1945, we wanted to have a fresh start and it seemed to us more likely to be found in the clear air of the mountains of North Wales among the realities of an old tradition of farming.

In two previous books (*Place of Stones* and *Hill Farm Story*) I have already told the story of our first twenty years at Carneddi—how we learnt the hard way, by trial and error, how we had no electricity or piped water, how our nearest and only shop was three-quarters of a mile away down a steep track and across muddy fields. But luck was on our side at the time. Land was cheap and costs were low, while farm products were modestly remunerative. We had the best possible neighbours on each side of the farm. I expect they had a good many laughs at our expense but they were unfailingly kind and unstintingly generous with their advice and loans of equipment and machinery. Without them, we should have fared badly.

We picked up bits of farming lore, we learnt to shepherd sheep, milk cows, make hay and grow crops on our few barren acres and, surprisingly, we managed to make ends meet, though it was always only just. I found that there was great scope for my love of animals and that I was a pretty good stockman most of the time. The beauty of the hills in their seasonal raiment and moods of weather was a constant inspiration and I found even a certain excitement and challenge in the hand-to-mouth state of our finances. It put lines on my face and grey in my hair but there was also a touch of exhilaration about the struggle.

In the earlier books I have tried to describe the small and beautiful hill farm where we lived and where we grazed our flock, reared a few pedigree cattle and ponies, kept hens and turkeys and grew a few crops. I have told how Paul and I married in 1960 and set up house in the ancient cottage of Tŷ Mawr, a few hundred yards from Carneddi. In 1966 our daughter, Ann, was born. After that the story begins anew.

It was wonderful to have a baby daughter, but now the happiness of our state was threatened by illness. A certain numbness which I had suffered was diagnosed as being caused by multiple sclerosis, a progressive and incurable illness. This was a shattering discovery and passed a sentence on me which I found very hard to accept. It could be the end of twenty years of work and dreams. An invalid wife and mother is a poor deal for husband and child, and Carneddi is not situated in wheelchair country. How could I shepherd the sheep if I could not walk? I kept on putting one foot in front of the other, and I must have been one of the lucky ones able to do it. Ann was just three weeks old when I lost the sight of one eye. I looked through the other and wept in secret at the prospect of becoming blind. Soon I decided that it was better to do something useful and feel ill than to do nothing and still feel ill. Then, mercifully, I became much better and most of the sight returned to the affected eye. It was the beginning of a remission that lasted for a couple of years. There were to be other exacerbations but I did my best not to let them affect our lives and we all carried on as well as we could.

A joyful event which made me forget the trials of illness was the birth of our son, John Piers, on 29 October 1968. He was a big boy with a wide mouth and large appetite. A friend of ours, Brenda, came to stay with us for a fortnight after John's birth. She was a district midwife and had married Brian, the old friend who had produced the illustrations for my previous books. Her presence was invaluable and it took all the worry and hard work out of having a new baby. I have pleasant recollections of watching John on Brenda's knee. In those days, he looked a little clerical in his white nightie. Often he would splutter after taking his complementary bottle too fast. Brenda would pat him on the back and inquire: 'Were there bones in it?' Those were happy days.

There was to be another addition to the family in four years' time. Paul's widowed sister, Jean, known to the family as Beenie, came to live with us. She came in the first place in answer to a cry for help. We had all suffered from a particularly unpleasant and virulent type of chicken-pox and were slow to recover. We had no farm pupil that year and we were getting so far behind with the farm work that we knew we should never catch up again that season. Beenie rolled up her sleeves and set us to rights. She loved farms, knew how to milk and was good with livestock as well as being an experienced and talented home-maker. After that we couldn't bear to part with her. She was the answer to a prayer.

We missed Fred badly when she went to Australia and married an Australian. When Betty, our hard-working helper and honorary Stud Groom, left us to get married, we missed her very much too. The pre-college students, who sometimes came to us for practical experience, the 'Serfs' as we called them, were often excellent, but they were too young to take much responsibility and too inexperienced to be of more than limited help. Now Beenie's presence added greatly to our family life and gave us extra freedom.

In the autumn of 1972, Brenda and Brian invited us to stay with them in Harrogate. Beenie said she could manage the farm for a few days with the help of Jenny, a very capable ex-Serf who was coming to stay. Paul and I were delighted; we hadn't had a proper holiday for years. I loved taking the children to see new sights. The sea, mountains, green fields and all our animals were everyday life to them and they found trains, buses, motorways and the trappings of civilization fascinating. They commented on the luxuries of central heating, fitted carpets and clean pavements for walking, because they were not accustomed to any of them. They looked forward to seeing Auntie Brenda's new baby daughter, Helen Louise.

We had an excellent holiday. I had not been to Yorkshire before and was charmed by the countryside. The weather was good and the trees were beautiful in their autumn colours. We went to Fountains Abbey, Bolton Abbey and the Stryd. We went to Grassington and crossed the River Wharfe on stepping stones. We went to Thorpe, where Paul was born, and then went down the Stump Cross caverns. We rowed on the River Nidd. We went to York and visited the excellent museum. The children rode on the escalator in Marks & Spencer's, the first they had ever seen. Twice we went to Knaresborough Zoo, and this is where the story should really begin.

# 2 Zoo Visit

Knaresborough Zoo was a small one, occupying a corner of the grounds of the Conygham Estate. Paul and I and the children went there one sunny morning, taking a picnic with us. When we arrived at the gates, we noticed a very tame golden pheasant rushing up and down in agitation, trying to get back inside the zoo. It obviously preferred its home there to the freedom of the surrounding parkland.

Inside there was much to see. There was the usual collection of zoo animals and birds, with the exception of some of the bigger ones such as elephant, giraffe, rhinoceros and hippo, but there were other special exhibits. Here we saw the largest lion then in captivity, a magnificent beast named Simba which had appeared on television several times and which had taken a star part in the film *Cleopatra*. Mr Nyoka, the proprietor of the zoo, appeared to be on friendly terms with the lion for he was in the cage with it when we arrived, pouring milk into its mouth from a milk-bottle. The lion was drinking the milk with evident enjoyment. Ann and John watched in astonishment. They had never seen anything like that before—man, lion and milk-bottle all in the cage together. The zoo was almost deserted that morning and Mr Nyoka had time to chat. Through the bars, we discussed the habits of lions in general and Simba in particular. It was clear that Mr Nyoka knew his lions and that he was a fine showman.

We also saw a huge reticulated python which, the notice said, was the longest snake in captivity, measuring 27 ft 4 in in length. These were exceptional creatures but there was something homely about the zoo which appealed to me. Except in the case of the more dangerous animals, the barriers between inmates and public were modest. You could approach the animals closely and have a really good look at them. There was the air of an exotic farm about the place and the animals looked well. We watched Mr Nyoka and his wife going about the business of feeding them, moving in and out of the cages with a friendly word to the inmates, as casual and relaxed as people feeding household pets.

A friendly nilgai, the large African antelope, leaned on its low fence and seemed to enjoy having its face rubbed while eating leaves from our hands. It shared its pen with a small bullock and several goats. Indeed, the smell of billy-goat pervaded the zoo. I noticed that the vultures were picking at the whole carcase of a still-born calf, and I thought that in a more sophisticated zoo, the meat would have been offered in anonymous chunks to spare the sensibilities of the visiting public. Peacocks and guinea-fowl pecked in the pathways. A husky dog was chained to a tree nearby and wagged its tail at our approach.

There was a pen of seven or eight llamas in the middle of the zoo, with a shed and a few tall trees in it. The pen was surrounded by diamond wire-netting of about shoulder height and the ground was covered with gravel. One or two of the llamas came and craned over the fence, obviously interested in us. The children soon discovered that llamas like leaves and ran about collecting those which had fallen in the walks between the pens. Paul and I stood and watched. The sunlight filtered through the trees. There were few other visitors at the zoo and it was very peaceful. Now and again the parrots screeched or there was a low growl from the black panthers to remind us that exotic fauna from other parts of the world inhabited this wooded corner of Yorkshire.

I was very interested to look at these llamas at close quarters. At first I thought there was something sheep-like about them, very large sheep indeed, with very long legs, very long necks and very long ears. But no, they were not sheep-like. Their elegant and dignified manner seemed unique and so was their expressive use of the ears to show emotion. They did not have cloven hooves but very small two-toed feet with a black claw or toe-nail on each toe. As they reached for the leaves, I saw that the upper lip was divided almost as far as the nostrils, making two little prehensile lips which manipulated the food. I knew that llamas were members of the camel family and were native to the High Andes in South America where they were used as pack animals. They were quite unlike any creature with which I had had dealings before. It was fascinating to watch them so closely. Of course I had seen llamas in zoos in the past, but never so near to me, only in the distance in a large paddock or an enclosure surrounded by water-filled moats.

Then, as I looked, the urge which had made me write: 'I want a llama' in my notebook all those years ago and which had lain buried for so long, suddenly came to the surface like a bubble bursting on a pool. All in a moment, like a revelation, came the feeling that this urge was not a ridiculous whim but a practical possibility. I knew nothing about llamas but, as I looked at the animals, I felt convinced that one could fit usefully into our hill farm, that indeed there was something very special about llamas which I had not known before. A small voice in my mind said very clearly and definitely: 'We'll have one of those sometime,' and a sensation of great pleasure and excitement came over me. I stood in a dream, looking at the animals and realizing that what I needed was a llama. 'We'll have one of those,' my mind kept repeating.

The children's supply of easily obtainable leaves for the llamas was coming to an end and they wanted to see something else.

'Come on,' said Paul. 'Let's have a look at the sea-lions.'

I came out of the dream and walked over to the pool with the others.

'How about having a llama ourselves?' I said.

Paul thought for a minute.

'We could,' he said, 'but we should probably have to have very good fencing to keep it in.'

It was a practical consideration. Paul was the person who struggled with our fencing problems, building the fallen walls and putting up netting and barbed-wire in an effort to contain our wild mountain ewes. Llamas were twice as tall as sheep and would probably need the sort of fencing used for deer, though neither of us knew if a llama could jump. But I could see that the idea appealed to him. He was a mountaineer. He found anything to do with mountains was of interest and a llama was a very specialized mountain animal.

'They certainly are beautiful animals,' he said.

Then I asked Ann and John if they would like to have a llama. 'Yes,' said Ann and 'Yes,' said John.

After we had seen all there was to see, we went back and told Brian and Brenda that we now fancied llama-keeping. They may have been surprised but they both thought it was a good idea.

Before our holiday was over, we paid another visit to the zoo. Again I took a long look at the llamas and again the feeling was just as strong that a llama would be right for us. The next day we went home, much refreshed after an excellent holiday and with much to think about. Now we set to work on our autumn tasks, lifting the potatoes, collecting bracken for bedding and gathering the ewe lambs ready to go away for wintering. The llama idea was still firmly fixed in my mind and we discussed it from time to time. Apart from the problems of keeping an animal which was quite outside our experience, there were two other nearly insurmountable difficulties—where did an ordinary private person buy a llama and how could we ever find the money for one? They were sure to be very expensive.

The first thing to do was to collect more information on llamas, I decided. I wrote to the Director-Secretary of the Chester Zoological Gardens, asking if there was any reason why a llama should not be kept on a hill farm. The reply was most helpful. No, wrote the Director-Secretary, there was no reason against keeping a llama in the same way as one might keep a pet horse or cow. They were domestic animals anyway but, he added, they had the nasty habit of spitting. We had heard about the spitting already and thought that it sounded less hazardous than the biting, kicking and goring of which our present livestock was capable.

'We could always get one of those notices that used to be on buses,' said Paul. '"Spitting Prohibited. Penalty £5."'

Among our many books on animals, there was hardly a mention of llamas. The only book which yielded much information was Volume Two of Oliver Goldsmith's *Animated Nature,* first published in 1774. Our copy was published in 1815 and had belonged to my great-uncle. Here we found four pages entitled: 'A Short History of the Lama'. It was fascinating reading. A llama, I read, 'resembles the camel, not only in its natural mildness, but in its aptitude for servitude, its moderation and its patience . . . like the camel, it serves to carry goods over places inaccessible to other beasts of burthen; like that, it is obedient to its driver, and often dies under, but never resists his cruelty.'

'Poor creatures,' I said. 'If we have one, we shall try to make up to it for the sufferings of its ancestors.'

I read that llamas 'inhabit the highest regions of the globe, and seem to require purer air than animals of lower situations are found to enjoy'.

'It's not very high here, but I should think our air is as pure as any in the British Isles,' I said.

I read that llamas in 'Potosí, and other provinces of Peru [Potosí is now in Bolivia], make the chief riches of the Indians and Spaniards who rear them: their flesh is excellent food; their hair, or rather wool, may be spun into beautiful clothing and they are capable, in the most rugged and dangerous ways, of carrying burthens not exceeding a hundred weight, with the greatest safety . . . they are, however, but feeble animals; and after four or five days labour, they are obliged to repose for a day or two. They are chiefly used in carrying the riches of the mines of Potosi; and we are told that there are above three hundred thousand of these animals in actual employ.'

Then followed a description of the llama's appearance.

'I'm sure it is not "cloven-footed like an ox",' I said, 'and I certainly haven't seen "a kind of spear-like appendage behind, which assists it in moving over precipices and rugged ways".'

'You must remember,' said Paul, 'that that was written two hundred years ago, and Oliver Goldsmith had probably never seen a llama.'

'Well, I hope this bit is true.' I read out: ' "It requires no care, nor no expense in providing for its sustenance; it is supplied with a warm covering, and therefore does not require to be housed; satisfied with vegetables and grass, it wants neither corn nor hay to subsist it; it is not less moderate in what it drinks, and exceeds even the camel in temperance."

'It sounds the ideal animal,' said Paul.

'Here's something about the spitting,' I said. ' "It is supplied by Nature with saliva in such large quantities, that it spits it out on every occasion; this saliva seems to be the only offensive weapon that the harmless creature has to testify its resentment . . . this fluid, which, though probably no way hurtful, the Indians are much afraid of. They say, that wherever it falls, it is of such an acrimonious nature, that it will either burn the skin, or cause dangerous eruptions." Hm, perhaps we shall have to find out what is the best ointment to apply to skin that has been burnt by spit of a very acrimonious nature!'

I found the llama idea fascinating. From time to time I discussed it with my mother. She was a little surprised at first. Indeed she had never before considered the likelihood of owning one, but soon she was convinced that a llama was not only a possibility but a necessity. That was one of the fine things about my mother—she took a very sane view of life.

I couldn't discuss llama-keeping with my father just yet. There were too many unanswered questions. He would worry about them and, perhaps, create problems in his own mind where there were none. The time to tell him was when all the details were worked out and a llama was soon to arrive. I had resorted to this method years ago, before the advent of the goats in Nottingham and my Alsatian dog. It worked smoothly. My father was a great animal-lover and I was fairly certain that he would be delighted with a llama when he saw it.

It was different with my sister, Mary. I could tell her about our llama plans and show her any llama photographs that I had found. Indeed llama plans and daydreams made a pleasant escape from her grim present. In the past, she had had a distinguished career as a teacher and had enjoyed life at Carneddi during the school holidays. Then she became ill. Though she went from specialist to specialist, the cause of her illness was not discovered until it was too late. It was found that she had a massive brain tumour which had been growing for many years. She went to Manchester for brain surgery, accompanied by our mother, and was there for the next eight months. Now she was home again. All that could be done for her at this late stage had been done. She was very ill. She could not read, write, walk or eat easily on account of pain, dizziness and sickness, though her considerable intelligence was as clear as ever. One of the few pleasures she could enjoy—as much as she could enjoy anything—was to take short walks with either Mother or me to support her. We would go slowly along one or other of the paths from Carneddi for a hundred yards or so, then find a convenient bank or rock to sit on and have a rest. We came to know every comfortable sitting-place within easy reach of the house. Here we would look at the beautiful views of the mountains which surrounded us and discuss pleasant topics, such as the garden or llama-keeping. Then we would go slowly home again, with none of Mary's problems solved though, perhaps, they had been a little alleviated for a short time for one more day.

# 3 The First Step

The New Year came. 1973 looked as though it would be a fairly difficult year, and it was. All our farm expenses were rising rapidly but the prices that we received for our livestock remained the same. With a growing family, it was more difficult than ever to make ends meet. Mary's illness was a constant source of worry, and at times we despaired for her. Also it saddened me to see my mother having to nurse her day and night. At the age of eighty, Mother should have been having an easy life. I did all I could to help but it was not enough.

The weather was good that summer. We had long sunny days that were pleasant for Mary's walks. We sheared, made silage and hay and lifted our potatoes, all in favourable weather. There was enough rain to keep the crops growing but not so much that outside work was difficult. The children and their friends, Amanda and Penny, had a craze for making camp fires that year. They boiled eggs and roasted potatoes in the embers for their supper. On most evenings a column of smoke rose from their fire down in the hollow below the cottage. There was great activity below and we could hear the sound of excited voices. The children were happy and filthy, and nobody got burnt on the camp fire.

We might have trials, difficulties and anxieties; hill farm life was not easy or secure, but it gave me immense satisfaction and pleasure to see children—ours and other people's—having the tune of their lives on the farm. I loved to see them running wild on our hillside, learning the ways of animals and the lore of the countryside, without ever knowing that they were learning, and getting rosy cheeks and becoming dirty, happy and healthily tired in the process.

Ann was seven now, a tall child with long fair hair. I felt at times that she was almost Ruth over again, she was so like me—in tastes as well as in looks. She loved the ponies and frequently rode Princess, one of the first two ponies we had bought in 1962. Although pony-breeding was not so profitable now, the ponies gave Ann so much pleasure that we considered them an indispensable part of the farm. Princess was now twenty-four years old, but she had a foal most years and was the best ride of them all. She was already broken-in for riding when we bought her and was an ideal child's pony, besides being our best brood mare. She was lively and spirited with a competent rider, slow and careful with a beginner. If her rider fell off, she would stop and look down at the child sympathetically, waiting till she was mounted again.

Amanda and Penny were also delighted by the ponies. They were a year or two older than Ann, but the three girls played happily together. John sometimes joined the girls but more often he tagged round with me; at times it was difficult for him to keep up with the big ones. At nearly five, he was a stocky child, big for his age, artistic, musical, something of a perfectionist and, we feared, a little deaf. He inhabited his own dream world most of the time. Perhaps he was a little like Paul, but mostly he was only like himself.

Then, in September of that year, my father died very suddenly. Paul, the children and I had gone to mid-Wales for a camping holiday but it was cut short after four days by the shocking news. We returned immediately to the sad business of bereavement, which had never affected me closely before. I was thankful that my father had not had to endure a long illness, but I felt stunned by our sudden loss.

After that, we carried on as one always does when farming. We missed my father's presence and help a great deal and, at every turn, found little tasks which he used to do and now were not done, also things he had made which were there to remind us of his enthusiastic interest in the farm.

My sister's health was deteriorating, Mother seemed to be living solely on her immense inner strength and I worried about them most of the time. My days were spent in hurrying up and down between Carneddi and Tŷ Mawr, doing a few jobs here and a few jobs there and noticing all the time the vast number of other jobs which were piling up, not done. Beenie was a tower of strength. I don't know how I would ever have managed without her, and Paul and I blessed the day she had come to live with us.

On most evenings, I would go to Carneddi after supper to sit with my mother for an hour or so before bedtime and try to lessen the emptiness left by my father's death. Mary went to bed early and usually had her last needs attended to by nine o'clock. I would hurry up night after night, gum boots squelching through the muddy patches on the path, now often wearing my father's old padded jacket, too big, inelegant but very warm. The path climbed up rocky steps between stone walls and then emerged to a green slope of field. Above this was the brow of the hill, just before Carneddi house and buildings came into sight. This green path was one of my favourite places. On the right, the ground dipped to the oak woods on our lower land, then to the forest and then down into the valley with the rocky ridge of Yr Arddu beyond. Ahead, the tops of the fir trees near Carneddi house towered above the rise, and behind them the rocky cliffs and heather of our higher enclosed land. To the left was the massive bulk of Moel Hebog. Sometimes the wind roared in the darkness and rain dashed against my face; at other times it was clear and peaceful, with the moon lighting up the surrounding mountains. Sometimes I walked into the edge of a cloud halfway up the path and the light from my torch was reflected back from a wall of mist, and the lights from our deep-litter henhouse made a diffused glow in the sky.

My mother and I were industrious in the evenings. She would be busy with her knitting or mending, and I was making a large rag rug, using old blankets and woolly garments cast off by members of the family. The effect was good and I enjoyed making it. Sometimes we would watch the television news and feel thankful that we were here on our own remote hillside and away from the turmoil of the outside world. Mostly we would chat over our work and drink our coffee. We discussed the happenings of the day and, perhaps, made pleasant plans for next season's work in the garden. Occasionally I would speculate on the possibility of having a llama.

Then, one evening, Mother said: 'If you can find a suitable animal, I'd like to buy a llama for you.'

I knew Mum was a specialist in making people happy, but the thought of her spending her meagre capital on such a purchase, especially when she was so beset by troubles and hard work and might have so many other uses for it, seemed wrong.

'No,' I said, 'you mustn't. You can't afford it. They are sure to be expensive—and whoever heard of an elderly lady buying a llama!'

'But I should like to,' said my mother. 'I should like to have one. We need something to cheer us up.'

We talked the matter over. Mother was quite sure that she wanted to buy the llama. She felt we needed something new and exciting to distract our minds from the sorrow and anxiety of Father's death and Mary's illness. If I really wanted a llama, she would get one for me if she could.

'If you did buy one,' I said, 'I could pay you back later when it had earned some money for us.'

'I don't want to be paid back,' said Mum.

'Well, we'll see. It would be an interesting subject to write about, and I should think the wool is quite valuable too. We could use it as a pack animal to carry things up to Clogwyn.'

The cottage at Clogwyn was perched halfway up the hillside with only a steep and stony path leading to it. No car or vehicle could get anywhere near and all supplies had to be carried there. We let it, furnished, as a holiday cottage to help to supplement our small income from farming. Llama transport would certainly add to the remote charm of the place.

When it was time for bed, I switched on my mother's stove and electric blanket, made her night-time drink, filled her hot water bottle, gave her pills and a goodnight kiss, and then hurried down the mountainside to Tŷ Mawr. Paul and Beenie were sitting by the fire when I reached the cottage.

'Mum says she will buy a llama for us,' I said.

'Good for Mum!'

'Isn't that splendid!'

We fell to discussing ways and means of finding one, but did not come to any conclusions. The whole idea still had a dreamlike quality, still seemed rather remote. There were sure to be dealers in zoo animals, but we had no contacts and it would be difficult and expensive to travel the country looking. But one door had opened for us—the means of finding the purchase price.

In the autumn, Brenda and Helen, now aged two, came to stay with us again. Before they returned to Yorkshire, Brenda suggested that she should take Ann, John and me back with her for a change of scene. The only holiday we had had the year before had been the short four days' camping in mid-Wales. It would be a pity for the children to have no holiday this year, and we accepted the invitation with enthusiasm. We had no farm student so Paul could not join us, but regretfully had to stay behind to help Beenie with the livestock.

One of my main objectives in Harrogate was another visit to Knaresborough Zoo. With my mother's offer behind me, I might be able to find out how one set about buying a llama.

We went back to the zoo on another sunny autumn day. Brian had lent us a camera and we had provided ourselves with a colour film to record our holiday for Paul, and particularly to get some pictures of the Knaresborough llamas. On this visit we found there were fewer llamas than there had been in 1972. Some of the handsome black and white ones were missing, but there were two young ones—a white, and one with a brownish body and a white neck. The llamas were just as beautiful and just as friendly as we remembered them. The children collected leaves again, to feed to the faces craning over the fence. The little white llama came forward with its elders and seemed very tame, but the brown one stood huffily in the background and ignored us. It was interesting to have a close look at young animals. I guessed that these might be two or three months old but, with no previous experience, it was hard to judge. They were most attractive animals and looked like large fluffy toys.

After a long look at the llamas, I set about plucking up my courage to do some fact-finding. Now was the time to have a chat with Mr Nyoka. I found him busy among the animal cages. He was a powerfully built man, with a slight limp. He wore a bush jacket and looked, unmistakably, the lion-trainer and White Hunter. I was thankful that he was so friendly and approachable, as I felt that inquiries about the purchase of a llama—particularly from a middle-aged woman in neat, town clothes with a handbag and well-shined shoes—might seem a little odd, but he appeared quite composed. We were soon chatting about the animals. I found he was a brilliant showman and raconteur. He told us that he was born into the circus, had never been to school and had gone on many safaris in Africa, collecting animals. Ann and John listened, enthralled. Their suspicion that school was unnecessary was now confirmed. To please them, Mr Nyoka fetched herrings for the sea-lions and let the creatures out of their enclosure. They came flopping and honking round our feet, while he tossed the fish to them. Then, when the feeding was over, he opened their gate and they flopped obediently back to their pool.

I found it fascinating to be in the company of someone who was so much at home with such a wide variety of
animals. Now I asked him about the llamas: were there any
problems with them? Did they keep healthy? No problems,
said Mr Nyoka, they were very easy animals to keep—but
it was clear that the big cats were his special favourites
and that he found llamas were a little dull in comparison.
I explained that I wanted to buy a young llama to keep on
the farm and asked if he thought that one by itself would be
happy, since by nature they were herd animals.

'It would be all right.' he said. 'It's the same with all
animals—if you give them plenty of fuss, you're their friend
and they aren't lonely.'

I asked Mr Nyoka if there were likely to be more young
llamas next year. Would he consider selling one to me? I
said that we would be very short of hay that winter and
needed time to get prepared but that we definitely wanted
one next year. Oh, yes, he said, he thought there would be
more young next spring. He would be quite willing to sell
one of them.

This rather took my breath away. It was all so easy. There
were no problems. I might have been shopping for half a
pound of tea. We left the zoo, having taken many photographs, feeling quite excited.

That evening, after the children were in bed, I telephoned
Paul to hear how everything was faring at Carneddi and to
tell him what had happened at the zoo. He was delighted to

hear that buying a llama was quite a simple business, with none of the difficulties that we had anticipated. He quite agreed that next summer was the time to take the plunge. It would give us a chance to get used to the idea.

The days of our holiday slipped past with pleasant outings and picnics for the children. Ann, Helen and John were great friends and Brenda was an ideal hostess. But I kept thinking about those young llamas at the zoo. Why wait until next year? I felt sure that Mr Nyoka was willing to sell one of them now. Certainly we were very short of hay, and fodder was appallingly dear, but would the situation be any better next year? A baby llama would not eat much. Why waste a whole year? I began to feel that we had an opportunity now which ought not to be missed. I began to feel a sense of urgency. Anything could happen in twelve months.

A couple of days before it was time to go home, I rang Paul again and we exchanged the news.

Then: 'How would it be if I tried to buy one of those young llamas now? It seems a pity to waste a year and I'm dying to have one. They're both females. I don't think they're weaned yet so we should have a few months' grace to get organized. Shall I try to arrange something with Mr Nyoka?'

'Have a go,' said Paul.

'OK, I'll see what I can do.'

The next day Brenda and Helen, Ann, John and I all trooped back to the zoo again. This time I was here to do business. We paid at the entrance and went along the gravel walk between the llama pen and the vulture cage. It was quite early and nobody else seemed to be at the zoo. The children knew their way round now and ran on ahead to see their favourite animals — Ann to the pony and the donkey, John to the lion, while little Helen zigzagged from one to another. We found Mr Nyoka by the panther cages. He stopped working and greeted us in a friendly way.

'I've been thinking about the llamas,' I said, 'and my husband and I have decided after all that we should like to buy a llama this year. Will you sell one of the two young ones?'

'Yes,' said Mr Nyoka. 'Which one would you like?'

I had already decided that the brown baby looked the better of the two. I liked its colour and it appeared to be in better condition than the other, but the tameness of the white one was very appealing. It was hard to decide.

By this time, Mrs Nyoka had joined us. Since there was no one about at the zoo at that early hour, she had let the little white llama and its mother, Grace, out of the enclosure for some exercise. The baby's name, I learnt, was June. Grace was delighted to have some freedom and was galumphing joyfully along the gravel walks, head held high, long skirts of wool flapping with every stride, and little June scuttling along behind. I watched, fascinated. It was an extraordinary sight and reminded me of a down-at-heel duchess, suddenly gone skittish, with scarves and skirts a-flying, and beads and bangles dangling.

'We can't sell June,' said Mrs Nyoka, 'I'm too fond of her.'

She told me that June had been a frail baby after a difficult delivery and had needed a good deal of nursing. In the process, Mrs Nyoka had become very attached to her. That left the brown one for me.

This was called Annie, after a visitor to the zoo who had witnessed her birth. Albert, the big white stud-llama, was her father, and her mother, Victoria, was also white, a large, matronly animal, dressed in shaggy shawls of wool.

'How much do you want for Annie?' I asked.

Mr Nyoka named the price. It was a lot of money, but it was in line with what I had already heard about llama prices and seemed, in fact, quite reasonable. This was no time for haggling and Mr Nyoka knew his business. I felt he was being very straight with me and that he was a good person to deal with. I took his hand in the Welsh manner of concluding a deal and said: 'We'll have her.'

We arranged that he should keep Annie for another three months till she was properly weaned and eating concentrates. Then we would come and fetch her. Mr Nyoka asked if I would be willing to make a small deposit. He said he often had people who wanted to buy an animal and who asked him to keep it for a while. Sometimes he heard no more and was left wondering whether he should sell the animal to another customer or go on keeping it. I wrote a cheque for £10 and my name and address on a bit of paper. We were now very nearly llama-owners. It was a great feeling.

That evening I telephoned to Paul again.

'I've bought the browny one,' I told him.

'Good,' said Paul.

# 4 The South American Beauty

Two days later we were back in Wales. Now I could tell Paul, my mother, Mary and Beenie all the exciting details of our llama-buying spree. Everyone looked with great attention at the photographs we had taken, and agreed that the little brown llama looked lovely and they were longing for her to arrive at Carneddi. No one much liked the name of Annie.

Our llama had been born in the previous August—the eleventh, Mr Nyoka thought—and we calculated that she would be weaned some time in February—that is, if llamas were anything like foals. We usually weaned the foals at about six months old but, of course, the genus of *lama* was very unlike the genus of *equus,* so we should have to be careful in making comparisons. However, horses and llamas have a similar gestation period, about eleven months, and so perhaps we could assume that a similar period of suckling was usual. Anyway, Mr Nyoka would know.

The weather stayed open until Christmas and the grass continued to grow. We were thankful. It gave the sheep a chance and eked out our small supplies of high-priced feedingstuffs. The summer of 1974 had been disastrously wet, not only in Wales but all over the country. The harvest had been poor everywhere, with the result that hay, straw and corn fetched almost prohibitive prices. The markets were flooded with sheep and cattle for which farmers had no winter food, and hill farmers were particularly hard hit. We had to sell six young ponies for give-away prices because we could not feed them, but we retained as much stock as we dared. If that winter had been a hard one, we should have fared badly indeed.

We celebrated Christmas in our usual traditional style with all the trimmings—a home-reared turkey and a home-grown Christmas tree and holly. Ann was thrilled with her good second-hand saddle from Grandma, and John with his steamroller that really steamed. Mary was too ill to enjoy anything.

Whatever snow and frost the year was likely to bring usually came after Christmas. With the holiday over, our thoughts began to turn towards the llama and we wondered if we should have trouble in crossing the Pennines to Harrogate to fetch it so early in the year. Carneddi faces the sea and the mild south-westerlies, which bring our prevailing weather, usually melt the snow quickly but conditions might be more severe in the east. We had heard nothing from Mr Nyoka. Paul had been a little sceptical all along and now began to wonder if the deal was really on. We had heard from other sources that llamas, particularly females, were just about unobtainable. Our easy purchase seemed almost too good to be true and we hoped that the little brown llama really was ours.

'Let's ring Brenda then,' I said, 'and ask her to go to the zoo and see how the land lies.'

Brenda obliged, and afterwards telephoned to say that our llama was fine; Mr Nyoka said that we could fetch it any time we wanted. A couple of days later we had a letter from him, confirming this and saying that the llama was weaned and had begun to eat concentrates. This was what we wanted to hear and we began to make arrangements for the trip.

Unfortunately, I seemed to be beginning a period of illness again. The early part of the year was often a bad time for me. My legs didn't work too well and any quick movement made me feel dizzy, but nothing must be allowed to interfere with such an important event as fetching the llama. I wasn't going to be left out of that: I would ignore my symptoms as best I could. I got a supply of ACTH and some hypodermic syringes from our doctor. ACTH was used for the treatment of multiple sclerosis patients and it seemed to shorten the period of exacerbation, reduce the after-effects and also made me feel less ill. Originally the district nurse had come faithfully up the mountainside two or three times a week to give the injections when needed. Now Paul gave them to me. He said he had injected hundreds of sheep in his time so why not his wife? This saved time and trouble for everyone.

Though we again had no farm pupil, Beenie said that she could manage on her own for a couple of nights if we all went on this exciting mission. We usually milked two or three cows to provide our own milk, home-made butter and cream, and we reared the calves and young stock. Then there were about a hundred laying hens, the ponies and the dogs, but Beenie said she could cope with them all for that short time.

The next problem was insurance. We were used to transporting ponies, but had no idea how a young llama would react when taken from its mother and put into a horse-box for the first time. A veterinary surgeon had told us that llamas have fragile legs, and indeed you could guess this was the case by looking at them. We felt, with such a valuable animal, that we ought to have some financial safeguard against big veterinary bills or possible disaster. An ex-army friend of Paul's, who worked for a big insurance company, tried to arrange a policy for us but the company was not the least bit interested in llamas. It did not know how to assess the risks involved and regretted that it could not help us.

'Oh well, we won't bother about insurance,' said Paul.

Neither he nor I were keen insurers.

'We'll trust to luck and common sense as we usually do.'

I agreed with him.

It was an overcast day at the end of January when we set off for Harrogate. There was no frost, wind or snow and the roads were dry. The children had cushions, coats and rugs in the back of the Land Rover so that they could curl up and sleep if they wanted, but of course they didn't. It was much too exciting to be going to Harrogate to fetch a llama. They found it fascinating to see and smell the oil refinery at Runcorn, to zoom along the motorway, to see the little back-to-back houses on the outskirts of Warrington and wonder how anyone could live in them.

'Nowhere to keep ponies and ride,' said Ann.

'Nowhere for the dogs to play,' said John.

'The people who live there do other things, I expect,' I told them.

We reached Harrogate in the late afternoon and were welcomed by Brian, Brenda and Helen. We spent the evening discussing our plans. The little llama would occupy only a small part of the two-pony trailer; Brian and Brenda had some beds to transport to the house in Nantmor which they had recently bought. We would need to partition off a small area in the front of the horse-box for the llama. This could be reached easily by the groom's door and would be warmer and safer than the end near the tail-board. The beds could go at the back. It was a pity that there was no way in which beds could be used for the llama's greater comfort, but it didn't seem that sort of animal.

The next day we went to the zoo to see Mr Nyoka and have another look at our llama. This was Paul's first sight of her and he was pleased with what he saw. She had grown a little since last October but not a great deal. The adult llamas were picking at the hay in their pen. There were no dead leaves to offer to them now. The day was again overcast, but dry and rather cold. Little Annie stood in the background, withdrawn and not eating. She did not know what was going to happen to her the next day or that she would soon be a Welsh mountain llama.

Mr Nyoka had a job or two to do in the panther house, so he took us with him. We went into the feeding passage behind the cages. Here there was no barrier to keep the public away from the wire. I could see John's face break into a big beam at his close proximity to the panthers. Mr Nyoka scratched the great black heads through the wire, but when a huge paddy paw came through the bars to within an inch or two of the children, he said: 'Keep back a little.' Ann and John stepped back quickly but they were not scared. They were enjoying every minute of being behind the scenes at the zoo.

A brilliantly coloured parrot lived in the back quarters of the panther house where it shrieked, whistled and swore incessantly. The spoiled pet of a private home, it had been donated to the zoo because of its noisiness. Mr Nyoka had had to banish it to its present position because he couldn't put up with it anywhere else. Indeed talking was difficult in its presence but Mr Nyoka managed to tell us that sometimes he took a panther or a lion for a walk round the park in the evenings. I should have liked to see him. I made some remark about lion-taming, but Mr Nyoka was quick to pick me up.

'You can't tame a lion,' he said, 'only train it.'

When his jobs were finished, he took us to the caravan where he and his wife and small son lived. The path to the caravan was not for public use and we dodged round the safety barriers to enter it. On one side was the diamond-mesh fence of the llamas' pen, on the other more diamond-mesh which enclosed some grass and two or three tall trees. As we passed by, with John and me last, a half-grown lion scooted out from behind one of the trees and pounced against the netting beside John.

'It fancies a nice fat boy for lunch,' I said.

John's face broke into a big grin. After a minute he said with satisfaction: 'I expect I'm the only boy in the school who has been pounced at by a lion.'

The Big Game stories were starting early.

The Nyokas' caravan was a large one and the furnishings were exotic. Striped African blankets covered the beds. There were African wood-carvings on the window ledges and photographs of Mr Nyoka with huge snakes and lions. One bed was draped with a tiger skin and there was a lion's head on the wall and necklaces of teeth. The children's attention was captured by all these fascinating objects. Mrs Nyoka made coffee for us.

All the way from Wales I had been keeping a firm hold on my handbag; it contained the purchase price of the llama in £10 notes. I was not used to carrying so much cash. Now was the time to hand it over. As I counted out the notes to Mr Nyoka, I couldn't help smiling. This was the first time we had bought a llama. Were we being quite mad, I wondered? What would the future bring? I couldn't answer these questions, but still I had the strong feeling that we were doing the right thing. I felt that there might be a kind of happy magic about llama-owning and that we were taking a rather important step.

Mr Nyoka gave me a receipt for all the £10 notes, we said good-bye for the time being and returned to Brenda's for lunch. In the afternoon, Paul went off to buy some chipboard to make the llama's travelling compartment. We would need to protect the beds and also we knew from experience that a nervous animal travelled better in a confined space. There was less room for it to jump about and risk injury and it would settle better. We were aware that we had no insurance. The children played together in the garden, skipping and riding Helen's tricycle.

Paul returned with his chipboard and set to work to make the movable partition. We should have to load the llama first and come back to collect the beds. Presently the partition was finished and we made preparations for an early start the next day.

In the morning the weather was still mild and fine and, after a quick breakfast, we drove off to the zoo. The moment had really come at last. I felt tremendously excited and I think Paul and the children did too. Paul drove the Land Rover and trailer into the grounds through the service entrance and positioned the trailer near the gate of the llama pen. Mr and Mrs Nyoka were ready for us.

Mrs Nyoka then let Grace and little June out for their morning galumph so that there would be two less animals to get in the way. Then she fetched several apples which she handed out to the adult llamas. There was a jealous milling round as each animal grabbed an apple. The apples were too big for them to bite easily and they rushed about, each with its own problem of how to reduce its apple to manageable proportions. As soon as the big male, Albert, and the mother, Victoria, were successfully gagged with an apple, Mrs Nyoka turned her attention to the little brown Annie who had not joined in the apple bonanza. Skilfully, Mrs Nyoka edged the young llama through the now open gate and out into the walk where Mr Nyoka, Paul, the children and I blocked all escape routes from the trailer ramp. Mrs Nyoka shut the gate and we all closed in a little. Annie stood uncertainly on her own, wondering what to do, but before she could do anything, Mr Nyoka had stepped quickly forward, gathered her up in his arms, and in a second she was in the trailer. There was a brief scuffle as Paul and Mr Nyoka hustled Annie up to the front and enclosed her in the small compartment. Then we shut up the ramp.

I felt jubilant. The loading had been done quickly and efficiently with no distress for the animals. It was good to work with people who knew what they were doing; even our children were quite experienced animal handlers. Nobody had made any mistakes. Now for the next stage.

It was time to be on the move. When a vehicle is travel-ling, the animal inside is preoccupied with the movement and keeping its balance. It is when the vehicle is stationary that the animal may start getting into trouble, so we quickly said good-bye to the Nyokas and set off. We stopped briefly at Brenda's to load the beds. Here I opened the groom's door an inch and applied my eye to the crack. In the gloom inside, I could see that the little animal was now sitting down, tall neck erect and ears forward. This was most satis-factory. We had been afraid that she might become frantic when parted from her mother. Everyone else had to have a quick peep but we were careful not to disturb her. We loaded the beds as quickly as we could. There must be no delay now and, without stopping for a coffee or more than a brief word with Brenda, we took to the road again.

After about five miles we stopped to see how the llama was faring. She was still seated with her legs neatly tucked out of sight beneath her wool. We were glad that she was taking care of these reputedly fragile members. It was amazing to see her seated in so calm a manner, with ears alertly forward. She reminded me of a little well-bred lady wrapped in a lacy shawl, unflappable in the face of ad-versity and accepting with dignity a situation about which she could do nothing. It was a tremendous relief to find that she was travelling so well. With the llama so composed, it was ridiculous for us to feel anxious about the journey, and our tenseness on her behalf relaxed. I found I was smiling as we drove along. We were all smiling. It was exhilarating to drive through towns and villages where people never gave a second glance at the shabby Land Rover and trailer passing through. They little knew that a South American beauty rode inside.

Halfway home we stopped for diesel at a filling station. After Paul had paid, he asked the girl to guess what was in the trailer.

'A pony?'

'No.'

'A donkey? A cow?'

'No, no,' we said and opened the door a crack for her to have a look inside.

'Good heavens!' cried the girl. 'What is it?'

We told her and, grinning happily, drove away.

Later we stopped for a picnic. Before we ate, I put a bowl of flaked maize and oats, a wisp of hay and a bowl of water in front of the llama so that she could picnic too. She held her head a little higher, half averting her face, but did not rise. It seemed to be a gesture of rejection, so I shut the door again and left her in peace. As the food and water were still untouched when we were ready to go, I removed the bowls and we set out on the last lap for home.

The early January dusk had come long before we rattled through Nantmor and ground our way up the steep lane to Tŷ Mawr. It was now quite dark. Paul pulled up on the parking place by our double gates. He could not get the heavy trailer across the field which separated Tŷ Mawr cottage from the lane. We had hoped to reach home in daylight and lead the llama over to the stable which we had prepared for her, but now we felt we could not risk it. It would be tricky to get her out of the small door and the beds would have to be unloaded before we could use the ramp. If the llama escaped us in the darkness, it might be difficult to catch her again; we had no idea how she would behave when we let her out. A few spots of rain had begun to fall. Perhaps it would be best to leave her where she was for the night. She must have become used to her present surroundings, and more handling and another change of scene might be upsetting for her after a day of travelling.

We left her and went across to the cottage. Here we found Beenie and a visiting friend, John Greaves who was usually known as J., having their supper. The dogs began to greet us with their usual enthusiasm, but stopped midway in the wagging and grinning and began to sniff deeply at our clothes. You could almost see the question marks. What on earth, they seemed to be asking, had their family been handling? With bulging eyes and great sniffs, they tried to interpret the strange messages on our hands and clothes. All through supper they kept coming back for another sniff, as though they could hardly believe their noses.

I put Ann and John to bed, and then J. and I went across the field for a final look at the llama. My dog, Taff, came too. He sniffed eagerly round the door as we opened it a crack to look at Annie. She appeared to be all right, still sitting there quietly with the food untouched before her. As we walked back over the field, it seemed wrong to be leaving her alone in the darkness and out of ear-shot of the cottage—this strange, exotic animal, far from anything she knew. But I didn't see what else we could do.

# 5 Home Again

We woke to the realization that something nice had happened—we had a llama. Paul, the children and I were soon across the field to see how she had survived the night. It was almost with trepidation that I opened the groom's door, but there was the llama and she seemed to be well. She was on her feet now. About half the flaked maize and oats in her bowl had gone. We returned to Tŷ Mawr for breakfast and to get ready to move the animal to her new quarters. We had no halter small enough to fit the little head and we thought that she might throttle herself with a collar; she was not used to being handled and we remembered the dynamic energy and wild struggles of untrained foals.

'She's as big as a small foal,' I said, 'but I don't think she's as strong.'

'I'll make a harness,' said Paul. He found a long piece of webbing from a pack-frame and constructed the harness from this. If she pulled, at least it would not strangle her. Then we all went across the field to where the trailer was parked, Paul and I, the children, Beenie and J. The weather had deteriorated and quite a strong south-westerly wind was blowing. There was a little rain in the wind and the clouds were low and dark. The light was poor; it was not a good day for photography but J. had brought his camera to record a scene which could never be repeated.

In daylight, we thought we could unload the llama through the groom's door; it was too wet to put the beds outside. We removed the food and water bowls. Then Paul stooped in through the low doorway and I shut the door behind him. He seemed to fix the harness without difficulty and then said, 'Open up.' I opened the door and the llama stood there, looking out. Paul gave her a push from behind and she leapt outwards. I caught her in my arms. She was not very heavy. Then, while Paul climbed out, I set her on the ground. I don't know what we were expecting her to do but she didn't do anything. Once in the open she seemed unwilling to move. She stood there, looking round and raising and lowering her head on its long neck, as though testing the ground before scanning the horizon. She made no attempt to spit. Although she was still a baby she had that extraordinary composure which she had shown throughout the long journey from Harrogate. Paul had a firm hold on the cord which was attached to the harness, in case she should make a sudden dash for freedom but she did not attempt it. She just stood there, her long white neck and fluffy golden-brown wool bright against the winter colours of her surroundings. She looked as though she alone were lit by a ray of sunlight in the sombre landscape—in contrast with the greyish, inconspicuous sheep grazing nearby.

We persuaded her to take a few steps and again she raised and lowered her neck. 'Pump-handling,' said Paul. Now, for the first time, we heard her voice. She made a little *mmm* sound, neither a bleat nor a moo, but a quiet musical note. It might be described as a sort of a toot. It was an unexpected and most attractive sound. The curator of a zoo had told us that llamas were mute, but this was one of many pieces of misinformation that we were to be given. Our llama was certainly far from being mute. Later, in my reading, I found the llama's voice likened to the sound of an Aeolian harp, a musical instrument played by the wind. Although I had never heard an Aeolian harp, I guessed this was an accurate description. In another book, the male llama's voice was described as 'singing'.

We crossed the field slowly, the little llama in her harness surrounded by admiring humans. Every few paces she stopped to raise and lower her head in that unfamiliar gesture and utter little toots. Besides this pump-handling, she displayed another strange movement. Every now and then she raised her neck upright and circled her head high in the air.

'What is she doing?' we wondered.

'Orbiting,' said Paul, but why she did it we could not tell.

We thought that we were experienced in animal behaviour but I began to realize that now we were faced with something quite new. For thirty years I had handled sheep, cattle, ponies, dogs, cats, goats and all sorts of poultry; I had brought up two children, had kept bees and even goldfish, caterpillars and a budgerigar, but the llama was like none of them. All of a sudden, I began to feel rather out of my depth. The llama certainly was different.

Now she walked along with her little fluffy tail upright. It was short like a goat's, but it was so thickly covered with gingery wool that it was hard to tell how she was holding it. The tail was up but it appeared to be crooked over backwards in a way which is not adopted by goats.

Finally we reached the barn and put the llama in a loose-box. With the harness off, she walked round it slowly, sniffing, pump-handling and orbiting every now and then. I put out a bucket of fresh water for her, a newly stuffed hay-net, another small feed of flaked maize and oats (Mrs Nyoka had told us that the Knaresborough llamas did not like bran) and a few chopped carrots and apples. The llama sniffed at my offerings, turned away and then lay down close to the back wall, front legs first, like a cow or a sheep, and then back legs. Then she sat up, very straight and symmetrical, with her swan-like neck erect, looking at us alertly. We looked back. Against the old stone walls of the stable, in the dull light of the winter's day, she looked very clean and bright. Her small head, long ears and long neck were pure white. Her body was mainly a goldish-brown, shading from mauvey-grey at the shoulders to gingery-brown at the tail. The brown colour met between her forelegs, like a mantle thrown over her back and fastened on the chest. Her front legs—though now we couldn't see them—were mainly white, though the grey colour extended to one of her knees. Both back legs were beige.

'I believe she's pretty thin under all that wool,' I said. 'I felt her back as we were coming across the field.'

'Let's weigh her,' said Paul.

'Good idea.'

Someone said: 'I don't much like "Annie" for a name.'

'No, I don't,' said Ann. 'We've got one Ann in the family already.'

'What about "Titicaca"?'

'Too long. "Titi" would be a silly name for a llama.'

'"Vilcabamba" — after the lost city of the Incas?'

'She would get called "Bamba" and that's too much like Walt Disney. A llama's more interesting than a fawn.'

In the end we resorted to a glossary of Quechua words, the language spoken by the Incas and the present-day Indians of Peru. The glossary was in a book that we had borrowed from the library. The book was an account of the Spanish Conquest of Peru. Even with this, it was difficult to find a name. Naturally the glossary was rather limited; the author had not been concerned with the naming of llamas. We found that *ñusta* meant 'princess' and that it was pronounced 'nyusta'. This seemed suitable for the daughter of Victoria and Albert, and indeed she had a real dignity about her. I didn't like the sound of the word much but everyone else did, so Annie became Ñusta.

We thought that we ought to leave her in peace for a while to get used to her new surroundings and perhaps eat, so we dispersed about our various farm jobs.

I was sorry that Brian was not in Wales then. I felt that we had reached an important moment in the history of our life at Carneddi and I wanted good photographs to illustrate it. Perhaps at some future date I would want to write a book about the llama and then I should need pictures. However, J. was a fine photographer and it was good to have him with us just then. Paul, also, took excellent photographs but he was usually so busy that his mind was on other things. Anyway I was fortunate to have photographers available. I hoped that J. and his friend, Miles Biggs, might someday feel inspired to make a ciné film about our new animal.

'It's not the inspiration that's lacking,' they said, 'it's the time and the money.'

They had already taken some 16 mm film of our ponies, but the costs involved were rising rapidly. They spent much of their spare time film-making. The movement of waves, water and wind, ripples and clouds, set to classical music, was the main feature of Wilderness Films and the results were superb.

We had first met Miles and J. about seven years before. One of the nice pay-offs of authorship is the people you meet. After my two books about the farm were published, a steady trickle of visitors arrived at Carneddi. It is perhaps unfair and less than accurate to describe them as 'fans'—a word that suggests uncritical enthusiasm—but we called them that for the sake of simplicity. Ann would find me and say in conspiratorial tones : 'Mummy, I think there are some fans coming,' by way of a warning. Fans were a part of her world; she sometimes received birthday cards and presents from people she had never met—her birth date was in one of my books. John was less fortunate because the reading public did not know of his existence.

The fans were mainly very congenial. They were not a random collection of people but a pre-selected group who had the qualities we appreciated: they were readers, they liked the country, animals and farming, they approved of personal endeavour and they were not daunted by ascending into the mountains, up steep narrow lanes, to find us.

Miles's parents were two such people. I found Paul talking to them when I came into the cottage one September afternoon in 1968. It was a month or so before John was born. At that time I did not realize that they were to be anything more than two of the usual nice fans who came, met us, talked and went away again. In the course of conversation they mentioned that their son, Miles, and a friend were filming in the district. We were interested and asked if we could meet them. A few days later Miles and J. arrived. Neither of them had read my books and they were slightly surprised to be summoned to the home of an unknown authoress and her husband, but they came.

So began a rewarding friendship. Now J. and Miles came from Liverpool quite frequently for weekends, to camp or later to sleep in the barn or the old converted hen-house where once I had gone to find the peace to write *Hill Farm*

*Story.* This was known as the Writing Hut, though it was now no longer used for the purpose. J. and Miles were interested in the farm and sometimes gave us valuable help. They were in need of fresh air and exercise, they said, after a working week in Liverpool, and muck-carting, scything and wall-building gave it to them.

Ann and John had no uncles, and every child is the better for having one. There is nobody who can quite take the place of a bachelor uncle. He is most valuable, as our children found. Now Miles and J. stepped into the role of honorary uncles with great skill. They read bed-time stories more interestingly than did parents. They became horses, hippopotami, elephants or Red Indians whenever the occasion demanded, emerging from the fray dishevelled and exhausted, but having given most valuable service. In return, Ann and John loved them.

In the afternoon of the day that we unloaded the llama, the light was so poor that J. was unable to take more photographs. Photofloods were needed in the dark stable. These, of course, were not available and J. was not a flash photographer. We decided to weigh and measure the llama instead. I fetched the bathroom scales and placed them on a flat slab in the stable floor. Then Paul put one arm round the llama's haunches and the other round her chest. As her feet came off the ground, she struggled wildly but then became still as he stepped on to the scales. She held her head high, ears back and an expression of alarm on her face, but she did not struggle again. The needle of the scales swung back and forth and it was difficult to make a reading. I crouched down, brushed back pieces of straw so that I could see better and, as the needle settled, was able to get the approximate figure. Then Paul set Ñusta down again. She retreated huffily to the far corner of the stable and turned her back on us. Next, Paul weighed himself and, by subtracting the second reading from the first, we discovered that the llama weighed about 4 st 6 lb. Whether this was a satisfactory weight for a llama of her age we did not know. By coincidence, six-year-old John weighed exactly the same amount. He was quite well-built but the llama felt extremely thin, with a knife-edged backbone and prominent ribs. I felt a little worried about this but, perhaps, members of the camel family were naturally bony. Next we measured Ñusta. She was 2 ft 9 in at the shoulder and over 4 ft to the top of her head.

During the day she seemed to have eaten none of the food I had put out for her.

'I expect she needs more time to settle down,' said Paul. 'Let's leave her in peace and give her a chance to eat.' So we did.

Last thing that night I went to see her again. She was still sitting quietly in her corner and the food was still untouched.

'You must eat,' I told her, 'if you are going to grow big.'

But she just looked at me with her enormous eyes.

# 6 An Anxious Time

The next morning, when I went out to the stable, I found Ñusta sitting in her usual corner, looking just as beautiful and just as exotic as before, but the food was still untouched. I removed all of it except the hay-net and put fresh water in the bucket. Throughout the day I tried to tempt her with various sorts of food. I tried meadow hay instead of seed hay, pony nuts, crushed oats and flaked maize—together and singly, damp or dry—twigs from various trees, warm milk and I even resorted to offering porridge oats and corn-flakes. All the food was rejected except for about a table-spoonful of dry flaked maize. She lipped it up slowly and daintily and did not finish the small handful I had put before her. After the greedy feeding manners of our cattle, ponies and dogs, which devoured everything they could get, her reluctance to eat seemed very strange.

Later in the day my mother walked down from Carneddi to see the llama for the first time. She was entranced by the new animal and delighted that it had arrived safely. After I had escorted Mum home again, we took Ñusta for a walk round the field in front of the cottage. She behaved well and did not try to fight against the harness and the lead. There were many stops for orbiting and pump-handling, but she did not try to graze. Now and then she assayed a short galumph with whoever was leading her running beside. There was great competition between Ann and John about whose turn it was to lead the llama. We were all enchanted by the animal. The more we gazed at her, the more of a fairy-tale creature she appeared to be.

But I wished she would eat. I went to bed that night feeling rather worried.

The next day was a Monday and the children went to school.

It was a beautiful, spring-like morning such as sometimes comes early in the year, though snow and frost may follow later. So far we had had an exceptionally mild winter. The snowdrops and crocuses were in full flower at Carneddi and the carpets of daffodils at Tŷ Mawr were already showing yellow.

It was Ñusta's third day with us and still she was not eating more than a few crumbs of flaked maize and occasionally a single strand of hay. I had not seen her drink and there was no sign of droppings in her pen. I knew that camels could go without food and water for many days, but was this the case with llamas? To our knowledge she had eaten next to nothing for four days. There was no reason for her to fast when there was food in abundance for her.

I should not have felt so worried if she had been plump and solid, but she was pitifully thin under the golden fluff. I knew how rapidly a young animal could decline when it reached a certain stage of debility, and how the decline was often irreversible. Why was she so thin, having come straight from her mother, I wondered? Perhaps she was burdened with intestinal parasites, but how could we tell if she made no droppings to be examined by the laboratory? I felt that she had few reserves to draw on in a time of stress, and I didn't doubt that now it was a time of stress, for all her calm behaviour. Llamas are naturally herd animals and here she was in strange surroundings with strange people and no other llamas to keep her company. Perhaps we had been foolish not to budget for a pair of animals. It was clear that she was pining.

Mr Nyoka had said that she would be happy if we made a pet of her. This we were now trying to do. Paul, I and the children had been taking turns to sit with her in the straw, stroking the soft fur of her neck and talking to her. She had been used to crowds of people on the other side of the fence at the zoo, had indeed lived with a lion only a few yards away in the next enclosure, but she was not accustomed to close contact with people. At first she flinched away from our touch, raised her neck upright and put her ears back. After a time she seemed more relaxed and, when we entered the stable, we were greeted by a little *mmm* sound.

Paul and I stood watching her. Neither of us felt very happy. Ñusta did not seem to be ill in any way but she just didn't eat. She regarded us steadily from her huge eyes, and her ears were alertly forward. Was it my imagination that made her look a little smaller and a little more frail than she had the day before? Was there a hint of despair in her quiet attitude of resignation? Could there be a danger that our little llama might fade quietly away until there was nothing left of her but a puff of wool and slender ivory bones?

Paul's next words did nothing to reassure me.

'I shall get some alum and saltpetre,' he said, 'then at least we should have the skin.'

He had recently been given the recipe for curing skins, using those ingredients, and was longing to try it out—but not, oh not on the llama.

'That mustn't happen,' I said with determination.

In the past I had nursed many animals which were gravely ill and some of them had pulled through against all expectations, saved almost, I felt, by sheer will-power. Of course modern drugs were great life-savers but, even if they were used, animals could die if one's attention wavered, even slightly. Sheep are particularly difficult patients. They suffer from a great variety of diseases and seem to give up the ghost very easily. Pet sheep make better patients than untamed ones. They seem to have more of a will to live, whereas the wild ones, when life becomes too burdensome, seem to die in self-defence. We retrieved Mooey, Ann's pet lamb, from the brink of death one year, and my attention never wavered for a moment in the struggle to save her life, though it lasted for weeks. She is lucky to be alive now. There have been other cases, less successful, where the patient lived for days, and then, perhaps overconfident that a cure was going to be effected, I slackened my efforts and, of a sudden, the animal was dead. Of course there have been times when the situation was hopeless, like the death from grass sickness of one of our pretty fillies. Grass sickness has nothing to do with grass; it is a nerve disorder of unknown origin and there is no cure for it. Cases of bracken poisoning, too, are incurable once the symptoms appear. Over the years we have lost a few calves from it, but I know that deaths from poisoning are almost always avoidable. They have occurred at times when my attention has wandered from obeying all the rules all the time.

On the whole we have been lucky and have had very few deaths among the larger farm animals. We have never regarded animals as expendable or mass-produced. If one does die, we mourn it and sometimes can say: 'It shouldn't have happened. We weren't taking enough care.'

One thing the llama wasn't going to be short of was care; I was determined about that. I must turn my whole attention on to the problem of getting her to eat and to being settled and happy in her new home, even without the company of another llama.

'Do you think vitamin $B_{12}$ would work for her?' I asked Paul.

'It's worth a try.'

In the past we had found that an injection of this vitamin would work like magic for an animal that was in poor condition and would not eat well. It was specifically for loss of appetite and debility but I had heard that it was sometimes given to horses before a race to give them extra pep—also, in a human context, for a hangover and *delirium tremens*. Whatever it did, it was marvellous stuff.

Next I telephoned the veterinary surgeon who attended the animals at the Mountain Zoo, Colwyn Bay. I had never met him but he had already given advice to me, over the telephone and for no fee, on the subject of llamas. If he thought me eccentric, he never said so, and was extremely courteous and helpful. Now he said that an injection of vitamin $B_{12}$ was likely to be beneficial. He advised us to wait till our llama was more settled before we gave treatment for possible worms. I was thankful for that. With Ñusta eating so poorly, she certainly wasn't going to take medicine mixed with her feed, and I didn't fancy trying to thrust pills or pour a drench down her very long throat. Cattle are easy enough to dose. They seem to have huge throats and anything gurgles down quite merrily. Ponies are more difficult and seem to be able to hold the medicine in the back of their mouths almost indefinitely. It is difficult to get them to swallow and, just when you think it has all gone down and your arms are aching with holding the animal's head in the air, the precious dose comes slobbering out again, often into your cuffs and up to your elbows. Llamas, I thought, might be even more difficult, with their special powers of spitting.

The next thing was to inject the vitamin $B_{12}$, so I boiled up the hypodermic syringe. Paul came along to help. Now, surely, was the time we should be spat upon. Paul held the llama while I tried to find an injection site. There didn't seem to be anywhere to plunge the needle in. Under the close wool there seemed to be only bones. I was used to injecting thin sheep, though I never liked doing it, and this was much worse. The llama was precious and I was sorry to give her any more unpleasant experiences, but it had to be done. I did it as expeditiously as I could. Ñusta flinched as the needle went in but Paul kept a firm hold on her. Then it was over; there had been no spitting.

'You are a good girl,' we said. 'You'll be better now,' and we hoped that indeed our words were true.

I fetched a small handful of flaked maize and put it in her bowl.

'It's all very well Oliver Goldsmith's saying that you "want neither hay or corn to subsist" you,' I grumbled. 'You are not even "satisfied with vegetables and grass". You turn up your nose at the vegetables and there isn't any grass at this time of year. You won't survive if you don't eat.'

As I said this I realized that I was wrong—there *was* some grass. Inside the fencing that enclosed the small vegetable garden at Tŷ Mawr, not yet dug for the spring planting, grassweeds flourished out of reach of the hungry sheep. I felt it would be better for Ñusta if she could roam round in the open and choose her own food. I was quite happy to sacrifice the few remaining cabbages, indeed would be delighted to see the llama eating them. Paul agreed that it was a good idea so we led Ñusta to the garden and let her go.

It was fascinating to see such a beautiful animal suddenly blooming in the little vegetable plot. We all watched eagerly to see what she would do and to be ready in case she tried to jump the netting, which now sagged a little with age. The llama slowly inspected her new domain, sniffing at this and that, going a few steps at a time and, every now and then, pausing to revolve her head—always from right to left, we noticed—in that strange gesture of orbiting.

'I wonder why she does that,' I said. 'Do you think she has something the matter with her ears, or is it because she has got such a lovely long neck and just likes to use it?'

'I think it is some sort of displacement activity,' said Paul, well up in his zoology.

'Maybe, but what activity is it displacing?'

That we did not know.

The llama quite ignored the tufts of new spring grass and the succulent cabbages, but she found some withered pea-haulms still clinging to last year's pea-sticks and began to pick at them. The sticks themselves also interested her and she chewed off some of the smaller dry twigs—not very nourishing, we felt—but it was a relief to see her eating anything. Perhaps the sticks would provide roughage. We did not like to leave her unattended because the wire fence might be a danger if she tried to get out. It had not been erected to enclose a llama.

'Perhaps she would be safer if we put her in Mum's garden,' I said.

The Carneddi vegetable garden occupied a small field of about a third of an acre below the house. This was enclosed by good stone walls. It had been a grass field when we had come to live at Carneddi thirty years earlier. In our second year at the farm our neighbour, William Owen, had ploughed it for us, using his plough and our horse. The field was steeply sloping and it was possible to plough it downhill only. The share did not penetrate more than two or three inches, but it was enough to turn most of the turf and it gave us a start. After that we dug and manured and carted off stones and boulders by the barrow-load until it was a good garden where soft fruit and vegetables could grow in abundance.

For the past ten years it had been much neglected. My mother and I, the gardeners of the family, had had more work than we could do—I with two small children, farm work and fluctuating health, Mother with an invalid daughter to nurse and the problems of advancing age. Recently we had managed to cultivate only the top half of the garden and the lower part had reverted to wilderness. In summer, bracken and nettles grew shoulder-high among the blackcurrant bushes and raspberry canes. The bracken had all died down now for the winter so there was nothing to harm the llama and nothing for her to damage, and the good stone walls would make a safe enclosure.

Moving very slowly, I was able to approach the llama again and to clip the cord on to her harness. Then, with Paul and Beenie in attendance and J. taking photographs, we led her up to Carneddi. It was a slow progress and the llama had to be coaxed along, but it was a historic moment: we were leading a llama where llama had never been led before. She moved with a graceful stateliness, blue sky overhead, February sunshine glittering on her golden wool and the beautiful backdrop of the mountains all around.

'Do you mind having a llama in your veg garden?' I asked my mother.

'I should be delighted,' said Mum.

So we put Ñusta in and let her go, hoping that she would find something to eat which would do her good. Mother and Mary stood on the upper terrace of the flower garden to look down on the little South American beauty that had appeared so strangely in their wintry garden.

We stayed for a while to see what the llama would eat but she didn't begin to munch the tufts of grass as we had hoped she would. Again she only picked at a few dry twigs and bits of dead fern, and nibbled at the moss and lichen on the stone walls.

As the sun was setting in the afternoon, I led her down again to Tŷ Mawr. She had had an eventful day. I hoped she would be the better for our efforts.

# 7 Settling In

The next day was bright and sunny again, though the weather was colder and the spring-like mildness of the past few days had left the air. Ñusta ate about a tablespoonful of flaked maize for her breakfast but would take no more. I noticed that there were a few little pellets of dung in the straw. This was a good sign, though there were pitifully few of them. At first glance the pellets looked like sheep droppings, but closer inspection revealed them to be elliptical with a little bottle-shaped neck at each end, definitely distinctive llama droppings and not like any I had seen before. I had read that the South American Indians used them for fuel but they didn't look very promising material to me. I couldn't imagine getting much of a fire with those. They were, however, healthy-looking and there was no visible sign of worms.

Paul's llama harness was a little complicated to attach. It tended to get caught in the wool so I decided to try a dog collar for leading Ñusta up to the garden. As before, our progress was slow but already the llama seemed more confident and relaxed. She seemed quite content with the collar and stepped along beside me, looking about with keen interest. She had huge eyes set in a prominent position in her small skull. They were big goggly eyes, fringed with heavy black lashes that might have been the envy of any movie queen. The irises were of a faded forget-me-not blue and, in spite of the rectangular pupils, reflecting unfathomable depths, her eyes had a strangely human look. Long grey and white hairs surrounded them, adding to the luxuriance of the lashes and enhancing the glamour of her appearance.

Her ears were busy when she walked—long, long semaphore ears, attached each side of the little hornless dome of her skull by what seemed to be universal couplings. They could move through almost three-quarters of a circle and seemed to work quite independently. They were covered with short white hair, very thick with the texture of plush. The skin below was dark grey and could be seen against the lie of the hair. Her ears had a slightly banana-shaped curve to them, spiralling a little before flaring out at the tips. I had seen other llamas, and pictures of llamas, with straight pricked ears and, in comparison, we thought that Ñusta's curved ones were particularly attractive. They were fringed by a little halo of long hairs. They looked like very beautiful and specialized pieces of equipment. We had read that the Peruvians sometimes earmarked their llamas for identification, just as we did the sheep, but it seemed a shame that such a perfect piece of evolution should ever be spoiled by marks.

I found that Ñusta was interested in the sheep. When we passed one, she seemed to grow taller, would sniff the air and gaze at it intently, ears right forward, and then make as if to follow it. The sheep, after a quick glance and sniff, were quite uninterested in the llama. I was sorry for her; she seemed to regard the ewes as sawn-off llamas with the wrong smell. It was a few weeks before she quite lost interest in them.

When we reached Carneddi, I let Ñusta go in the vegetable garden again. She had a tendency to hang round the gate, looking lost and peering through the wire, giving lonely little toots. I stood watching for a while. Presently she wandered off to begin picking about at the dead twigs and dry ferns. Occasionally she performed a very strange action which we had noticed on the previous day. Quickly she flicked one of the spindly back legs over the hock of the other and rapidly rubbed the outsides of the two back legs together. It was a most extraordinary action. I had never seen another animal do it and one would have thought it impossible if her legs had not had the insubstantial look of boiled sphaghetti. The movement reminded me of a housefly twiddling its back legs together.

In the late afternoon, I led Ñusta down to Tŷ Mawr again. I put her in the stable with a small feed of maize, fresh water and hay. Later in the evening, one of the vets from our local practice called to make a pregnancy diagnosis on two of our cows, Lovely and her daughter, Quix. Lovely was our oldest and best cow. She had given us ten splendid calves and generous bucketsful of milk night and morning from the poor rocky pasture. Her dam, Heather, now dead, had been equally good, but we were still waiting to breed another animal which had quite their excellence. Quix looked magnificent and would allow the children to ride her, but she didn't milk so well and had, so far, produced only bull calves. Sadie, another of Lovely's daughters, was proving obstinately difficult to get in calf and the vet was going to give her a hormone injection.

We had much satisfaction from our small herd of pedigree Welsh Blacks as well as the abundant milk, cream and homemade butter. In our isolated position and with such poor land, it would have been uneconomic to sell milk and we preferred a measure of self-sufficiency. All the cattle were descended from one heifer which we had bought with the farm for £10 in 1945, and we had no outside blood. We had graded them up to pedigree status by the use of pedigree bulls, mainly through artificial insemination. Now we were rearing calves of the tenth generation and there would have been many more generations if we had not preferred to keep the best old cows and sell the heifers.

Beenie had taken over the milking, dairy work and calf-rearing when she came to live with us. She was an expert herdswoman and took much pride and pleasure in her fine animals. The late Secretary of the Welsh Black Cattle Society once said that he was surprised that we managed to breed such good-quality stock when we had so few of them. I expect the fact that we were interested in them helped, but we, too, were very surprised when one of our heifers,

Carneddi Nerys, was the Champion in 1967 and sold for the then record price for the breed.

Nerys might never have gained fame if she had been of a different temperament. She might have spent all her life peacefully on the hillside of Carneddi, grazing among the bracken and the rocks in summer, keeping warm in the dark little cowshed in winter and feeding on home-grown hay and swedes, breeding good calves and filling the milk bucket twice a day. We might have kept her till the end of her life because she was the kind of cow we liked.

She was the daughter of Heather, and grew into a chunky little heifer with all the good points of the breed. She had the little bit of extra intelligence that always seems to go with good milking qualities. She was forever foraging on the bleak pastures, was the first to arrive at feeding time and was the one that could wriggle under or climb over the defences that we built to keep the cattle out of the growing hay. If there was something good to be had, Nerys was there to have it. We didn't mind this. It was when she extended her diet to less conventional foods that we felt she ought to go.

She and some other young stock were grazing on the Tŷ Mawr fields with the ponies. One morning we noticed that one of the ponies seemed to have rather a thin tail. Odd, we thought, what can have happened to it? The next day the pony's tail had completely disappeared. There was just a ragged bunch of hair left on the dock; the lovely heel-length tail had all gone. It was not until another pony had lost its tail that we saw that Nerys was the culprit. We found her just beginning to consume a third tail, plucking out the hairs while its owner stood placidly still. This odd appetite might have been the result of mineral deficiency but I doubted this because the animals were supplied, as usual, with mineral licks.

We moved Nerys to another field. Here there was a lamb which had had maggots and which we were keeping under observation. With no ponies' tails to eat, Nerys found that sheep's wool made a good substitute and began to denude the poor little animal of its ragged fleece. The victim didn't seem to mind, however.

Paul and I moved the lamb to a safer place, but we were worried; Nerys might get out in the night and eat our stallion's tail. She was an expert escaper and the stallion's tail was important. Without it we could not show him. At that time we were having success in the show ring against formidable competition with Carneddi Idris, our home-bred colt. A tail would take six months or a year to grow and we were not prepared to take the risk. Also, if Nerys were developing an abnormal appetite, she might be the kind of animal to pick up polythene bags, nails or barbed wire, the result of which might prove fatal. We decided to sell her.

We put Nerys in the next local sale for pedigree Welsh Blacks. She looked good and sold for a rather better price than the average for a yearling heifer. We heard she went to Scotland.

There was no more news of her for eighteen months. Then the telephone rang. It was a farming neighbour who had been to the annual show and sale of Welsh Black cattle in Perth. He wanted to congratulate us on the outstanding success of our heifer. We were astonished and delighted by the news that Nerys was the Champion. It was a pity that we had not been on the receiving end of that record price, but our herd prefix of 'Carneddi' had been given an advertisement. If we had kept her, Nerys would never have reached the required size and fatness for a Champion on our poor land but would have lived her life out in obscurity on the hillside. Did we, perhaps, harbour other potential champions? It was strange that a predilection for ponies' tails had caused the brief spotlight of fame to illumine our heifer's career.

When the vet had finished his work, we were pleased to hear that both Lovely and Quix were indeed in calf. Sadie had her injection and we hoped it would give results.

Then I said: 'We've an animal in the next shed which may be a patient of yours someday. Come and have a look at it.'

I led the way to the stable. The vet followed me in.

'Good heavens,' he said, 'it's a llama! Where did you get it?'

I told him. Over the following months we were to find that this was the question most frequently asked. Indeed it was usually the first question—'Where did you get it?'— as though the animal's extraordinary appearance and our reasons for keeping it might be explained by its place of origin.

The vet said that he had no experience of llamas but he was most interested in the animal. He agreed that she was extremely thin but could suggest nothing more for her welfare than the steps we had already taken.

Later that night I went out to have a last look at Ñusta. She was sitting quietly in her corner in the neat tea-cosy position. She gave a little *mmm* when I entered but did not get up. Then I noticed that she was chewing her cud. Oh, joy! This was a sight I had been longing to see. To the owner of any ruminant, the sight of the animal chewing its cud is a definite assurance that it is well and contented. Later on, I read that llamas are not classified as ruminants though, like all the camel family, they chew the cud and have a rumen, which is the first of their four stomachs. Some experts consider that the llama's third stomach, the omasum, is not sufficiently differentiated anatomically to form a separate compartment. It is thought that camels and llamas were derived from a pig-like animal in North America, quite independently from the true ruminants. This wouldn't have interested me at that moment; I was so delighted to see Ñusta sitting there, chewing contentedly. There was nothing much wrong with her if she did that.

I watched intently to make certain that I was not mistaken and that she was not just grinding her teeth. But there was no doubt about it. I saw a slight spasm in her abdomen and a second later a bulge was visibly travelling up her long neck. Then the bulge appeared in her cheek, and the steady rhythmical chewing began. She chewed rather more rapidly than a sheep and much faster than a cow. Her lower jaw moved quickly, left, right, left, right. Sheep and cattle, I had noticed, always chewed with a circular movement, the jaw moving clockwise for a while, then reversing and going anti-clockwise. So the llama was different in this respect too. She chewed fifty or sixty times and then swallowed. The bulge disappeared. There was a short pause and presently another spasm occurred and another bulge, the bolus, rippled up her neck. She sat there with a faraway look in her eyes, attending to the serious business of cud-chewing. I watched, fascinated and happy. I felt as though someone had given me £100. The vitamin $B_{12}$ injection had worked. The llama really must have found food in the old kitchen garden.

As I watched, I began to speculate about how our society would differ if humans had evolved as cud-chewers. Meals would doubtless be hurried affairs, with very large vegetable dishes to accommodate the required bulk. Every well-appointed house would have its ruminating room next to the dining-room, where people could retire to chew in silence. After-dinner conversation would be unknown. Patent medicines would proliferate, with four stomachs per person to medicate. There might be Civic Ruminatoriums in large towns and Cud Caves for hippy gatherings. The tenor of life would certainly be slower.

Soon I withdrew my thoughts from fantasy and hurried back to the cottage to tell Paul the good news about Ñusta.

That seemed to be the turning point. From then onwards the llama began to eat better. She would take two handfuls of flaked maize at a meal but she continued to be dainty in her eating habits. Her bowl had to be scrupulously clean or she wouldn't touch it. She always left a couple of teaspoonsful in the bottom for Miss Manners in the way that some Victorian children were taught to do. I found that she didn't like the hanging hay-net but would eat small quantities of hay if I put a handful on the ground near her favourite sitting corner.

Now we felt the tide had really turned. Excluding some terrible tragedy, the alum and saltpetre would never be needed. On 10 February Ñusta ate some maize out of my hand and I felt that this was another milestone in her progress. If an animal will come to you, contact has been established.

Each day, if the weather were fine, I would lead the llama up to the garden where she spent the day. Coming down in the afternoon, I found she would follow me without the lead and go willingly into her shed for food. This was another stage in her education. When she was free, it was interesting to notice that she was unwilling to cross the open field but made her way to the wall and walked along beside it. When she came to a gate, she would stand near to it but was most reluctant to go through. She still had the zoo mentality. Her six months of life had been spent in an enclosure and she expected the world to be bounded by fences. I noticed also that she was very awkward in negotiating the rocky path. She fussed about crossing the smallest obstacle and I had to coax her. It seemed odd that an animal which had evolved for mountain life over millions of years should show no agility because it came from a zoo. Now we found that she would stay happily near the cottage if we let her loose in the mornings. She seemed to like our company and showed no desire to escape.

One sunny morning at the end of February, when Ñusta was nearby, Paul said: 'Wouldn't it be nice to have the llama in the house?'

Of course I agreed. We had had ponies in the house and the pet lambs were difficult to keep out. Neither were very welcome. The lambs almost invariably left puddles and a trail of pellets, while the ponies were clumsy and it was sometimes difficult to back them out through the narrow hallway. From time to time the odd hen would potter in if the front door were left open, sometimes with a trail of chickens, and begin to search for crumbs on the kitchen floor. We had one that liked to lay its egg under the stairs. I quite enjoyed this invasion of animals but all too often I had to get the shovel and floor cloth afterwards. Once a heifer came indoors—at least she must have done though I never saw her. A small group of heifers were then grazing in the field in front of the cottage. One day I was rather alarmed to find the front door had been left open when I returned from Carneddi. Cows are clumsy creatures and will eat the most extraordinary things if they get the chance. On entering the cottage, it was a relief to find that everything indoors was neat and undisturbed. I thought I had got away with it that time. Everything was just as I had left it. Then I happened to glance towards the hearth rug. There, in the exact centre of it, was a neat, circular cow-pat, positive proof that we had had a bovine visitor which, except for this, had behaved in a very unbovine way.

Now we were thinking of adding a llama to our list of animal visitors. With a handful of maize, I lured Ñusta through our front door. She hesitated slightly but it was obvious that she was extremely curious and she followed me in with ears semaphoring. Her little leathery feet made a shuffling sound over the slate flagstones of the passage. In the kitchen she inspected everything with great interest and care. Her head reached well above the level of the working-tops and she could examine some of the shelves. She had just begun to eat small amounts of chopped swede, carrot and apple, so now I began to prepare some for her. She came to inspect. Then she began to eat the vegetables straight from the chopping-board. Paul and I watched the use of her lips with fascination. The upper lip was divided as far as the nostrils, like a rabbit's or hare's, but these two lips seemed very mobile and muscular. She could extend them to reach out, like two little fingers or two mini-trunks, and slide the pieces of carrot into her mouth. They were prehensile, a gun-metal grey covered with a fuzz of short silvery hairs, and they were immaculately clean.

When she had eaten a part of the chopped vegetable, the llama followed us into the sitting-room. We had the feeling that she liked being indoors but was slightly apprehensive of the strangeness. Also, she was still a little wary of us. We sat down so that we would not distract her and watched what she did. Again she was curious about the room and inspected its contents, moving very carefully so as to do no damage. The inspection over, Ñusta moved to the hearth rug and then, as though taking her cue from us, she positioned her feet close together and sat down. She sat calmly and with dignity, like an invited guest—as indeed she was. Paul and I grinned at each other; we had not expected this. We were amazed by the extreme domestication of our beautiful animal, and her delicacy, cleanliness and good manners. We felt we had an aristocrat in the family.

# 8 Waiting for the Snags

In the hopes of getting some guidance, we had made inquiries to try to find someone else who had a llama. Apart from zoos and circuses, there seemed to be no one.

'I wonder why people don't keep llamas,' I said. 'They are quite ordinary domestic animals in South America.'

'Perhaps people haven't thought of it,' said Paul.

'I can't think we're unique,' I said. 'There must be snags we haven't heard about.'

'We'll come across them in time,' said Paul, 'but I wonder what they are.'

From never having given a thought to llamas, he had now become a full-scale enthusiast. The more we found out about llamas, the more intriguing they seemed. Our golden South American beauty was developing immense charm and personality, and the history and development of llamas were, as far as we could glean any facts, fascinating. Recently we had been reading a book about the Spanish Conquest of the Incas—the book which had furnished us with the name, *ñusta*. It was a deplorable tale of cruelty, treachery and greed, but we found the occasional mention of the llamas most interesting. It was thought that the great Inca civilization could never have developed without this animal. Llamas were used as beasts of burden and also provided wool, milk and meat. The Incas inhabited the High Andes, and llamas could live and work on the high, dry *altiplano,* up to 17,000 ft where no other large mammal could survive. The Incas, though they never invented the arch, the wheel or the written word, were highly civilized. They were expert farmers, master masons, road builders and craftsmen in precious metals. Though the Inca—the god-king—and the ruling classes lived in great splendour, their flourishing socialist state, ruled by a benevolent, theocratic dictatorship, provided well for the people. In fantastic series of terraces down the mountainsides, maize and potatoes were grown, sufficient for all. There were large granaries and storehouses to provide plenty for the lean years. Huge herds of llamas grazed the mountains, and as many as 15,000 animals from the royal herds would be slaughtered at a single *chaco,* or round-up. The farmers practised good livestock husbandry and bred their llamas carefully.

Thus the Incas lived, high in the Andes in their stone-built cities or cultivating their terraced fields, transporting their produce by llama train along the precipitous paved roads, worshipping the sun in gold-lined temples where llamas and ears of maize, wrought in gold and silver, symbolized plenty and fertility, where the Inca, god-king, son of the sun, ruled over all. Here llamas also were venerated. They were used as sacrificial animals, the convolutions of their intestines were consulted by soothsayers and they were sometimes mummified along with the bodies of past

Incas. Here in the brilliant thin air, where the sun rose in the mornings and set in the evenings over the high dry slopes of the mountains, lived the sun-worshippers, the Incas.

Then, in 1532, came the Conquistadors, in the name of God, to bring religion to the heathens and perhaps to enrich themselves and the Church in doing so. The Conquistadors had metal spears, swords and armour and they had horses. The Incas had never seen a horse before and, at first, were terrified by these armed centaurs, but gradually they made friendly overtures to the strangers. The Spaniards accepted the friendship, but they wanted power. In November of that year, Atahualpa, the ruling Inca, was taken hostage by the Conquistadors. His freedom could be bought by a ransom, they said. If a room, which was 22 ft long by 17 ft wide, was filled to a depth of 8 ft with golden objects and again twice over with silver, he would be freed. The task was to be completed in two months, and it was completed. The temples were stripped of their heavy gold and silver treasures which were carried by endless llama trains until the room was filled three times over. Atahualpa's ransom had been paid.

Hernando Pizarro, leader of the adventurers, now decided that Atahualpa, free, would hinder further conquest. Treacherously, on 26 July 1533, the god-king was executed. More gold must come first; the souls could come later. And so eleven tons of gold, the beautifully wrought treasures of the temples of the sun, were fed into the furnaces of Cajamarca to be made into ingots to be shipped to Spain. On one day alone two hundred and twenty-five llama-loads of precious metal were brought from Cuzco.

So the land of the Incas was stripped and laid waste. The people fought bravely, retreating further and further into the high mountains, but with only primitive weapons, they were no match for the mounted horsemen with their armour, swords, spears and their fierce greed to enrich themselves. The Conquistadors stripped the temples, occupied the towns, broke open the granaries, slaughtered the llamas—which they called the Incas' 'cattle' and prized for the marrow bones—and took the people into slavery. In a few short years there was nothing left of that once flourishing civilization.

As we looked at our beautiful llama, it was fascinating to speculate whether any of her distant ancestors had trodden the royal road to Vilcabamba, the city of the Incas, had carried the treasures of Atahualpa's ransom to the furnaces of Cajamarca or been sacrificed on the altars of the sun. She had a glamour and dignity about her which seemed to hint at past glories, ancient mysteries and distant tragedy.

The days passed and Ñusta began to settle into family life. We waited expectantly for the snags of llama-keeping to appear but, so far, there were none. The llama's appetite was gradually improving and now she began to eat what we considered to be a modestly adequate amount. Her backbone was still like a knife-edge but, after the days of near-starvation, it was immensely satisfying to see her tucking into a dish of chopped swedes. She tested the food first, like a cat, and if it were to her liking, would begin to eat. It was a pleasure to see her with a mouthful of swede, a contented expression on her face, nose pointing slightly upwards the better to get the food to her back teeth, and munching loudly. No mother with a faddy child was ever more pleased to see her offspring eating than we were to see the llama doing likewise. Any marred or substandard bits were flicked aside by her selective lips. She never quite finished her food, always leaving a few pieces to remind us that she was still a dainty eater.

We knew nothing about a llama's dentition but we noticed that Ñusta had no top front teeth, only a dental pad like all the cud chewers. When she yawned, the mobile lips pouted forward at the beginning of the yawn and then stretched back at the end of it to reveal the dental pad which was narrow and dark grey. She seemed to have four incisors in the lower jaw, big broad teeth that didn't look much like milk teeth, though we supposed they must be. They were long and yellowish and sloped forward from her jaw like an old horse's teeth. Later she raised another pair, making six in the lower jaw. They were often partly visible when she was standing surveying the scene, giving her something of the air of a tall, toothy, upper-class lady as she looked about her. What her molars were like, we could not tell.

After her first successful visit to the cottage, she began to come in regularly. She spent the night in her loose-box in the stable where she had concentrates and hay, then, when let out in the morning, she would come tripping along to the front door. A white face would appear at the glass pane in the upper half of the door and a quiet voice would say: '*Mmm mmm.*' I had chopped vegetables ready on the draining-board, and in Ñusta would come to finish her breakfast in the kitchen. When she had had enough, she would retire to the sitting-room and seat herself on the mat where she would remain very quietly with a contemplative expression on her face. Sometimes she would chew her cud but more often she just sat there. Sometimes she would stretch her neck out along the floor in front of her and rest her head. She always lay in a perfectly symmetrical position with her legs invisible beneath the flowing hair. Probably she did this so that she could rise to hr feet again quickly. It was not until some months later, when her confidence was complete, that we saw her lolling sideways, forelegs still neatly tucked under, but the strange satyr's hind legs stretched out to one side.

The first time that Ñusta rolled on the sitting-room floor took us by surprise. It was extraordinary to see this large fluffy animal writhing on the floor, with its spindly legs flailing in the air. But what surprised us most was the quick glimpse of a tiny maiden udder with four teats. I don't know why we should have assumed that she had only two, like a sheep, goat or mare, but that's what we had thought. Now it was interesting to find that she had a mini-udder like a cow.

With the frequent comings and goings of the llama in the cottage, I was fully prepared to have to wield the shovel and the floor cloth. This was usually necessary when the current pet lamb managed to squeeze its way indoors, but again we were surprised. Llamas, we found, were much too polite to make messes in the house. If Ñusta wanted to go out, she went to the door and asked. Then she would hurry off to the llama lavatory which she had established under the plum trees, twenty yards from the cottage. As time went by she sited a series of middens in strategic places about the farm, all conveniently near but not on the paths she used. These she visited faithfully. We only once had an accident in the house and this was our fault. It happened before we had fully realized that it was possible for a baby llama to be completely and naturally house-trained—indeed that such an animal could ever be house-trained—and we disregarded her distress signals. We became wiser after that. It seemed marvellous to be able to invite such an exotic animal indoors and to know that there would be no need for the mop and bucket routine. Our feelings that she was a real lady were greatly reinforced.

We knew, however, from our reading and from hearsay that llamas had the definitely unladylike habit of spitting, 'a nasty habit' the Director-Secretary of the Chester Zoo had called it. So far Ñusta had not spat. We had injected her twice, we had picked her up to weigh her, we had led her round on a collar, enough, one might think, to make any-one spit but she had not. We began rather to long for her to spit to demonstrate all her llama-like qualities, but nothing happened.

She was the calmest animal imaginable. If the children dropped a box of Lego on the uncarpeted floor upstairs while she was sitting on the mat below, she did not react at all. If John fired his cap gun a few yards away from her, she just went on sitting. We were interested to read in an old army manual that camels were very good under gunfire. The same must be true of llamas. After being used to the nervous temperament of ponies, which quail at loud noises and which are prepared to see spooks round every corner, we found Ñusta's calm all the more unusual.

The dogs, after their first thrilled interest in her smell, accepted her presence in a matter-of-fact sort of way. There must have been nothing about her which appealed to their sheepdog instincts; I never saw them try to chase her, round her up or stare at her with the sheepdog's 'eye'. They skipped between her legs or dodged past her when this was the most convenient route, but always treated her as though she were none of their business. On the whole she ignored them. She only showed disapproval when they lay down too close to her or queued round her legs in the hopes of getting a stray crumb, when she was eating. Then her ears would go back and her expression said definitely: 'Push off.' Much later, I met a Chilean veterinary surgeon who asked if our llama was frightened of dogs. He told me that, in South America, there were cases of dogs killing llamas in the manner of sheep-worrying in this country. Fortunately this seemed to be one problem that wasn't going to bother us.

Although she was becoming very tame and domesticated, Ñusta didn't much like being touched, particularly by strangers. She would tolerate a little stroking on the neck but patting on the body caused her to withdraw with her ears back. She considered it a liberty. She would allow more latitude to me than anyone else. We found her aloofness rather tantalizing. She was beautiful to the touch, warm and soft as silk, with dry, clean fluff that left no trace of grease on the hand. We noticed that she had a faint but delicious smell, a smell—like all smells—difficult to describe. It was not quite like anything we had smelt before, but it might be described variously as musky, biscuity, nutty, like new plasticine or freshly baked pastry or, more prosaically, dry dog. Her stable had a strange, zooish, ammonia smell but this seldom clung to her hair. She had the sort of cuddly, irresistible attraction of a teddy bear, a teddy bear which was silky, centrally heated and animate. You longed to hug her but this was not allowed, except occasionally to selected people. She was quite happy for Paul or me to brush her, which we did most days to keep her tidy. The long fine hair picked up bits and debris easily from the stable or where she lay outside. She looked beautiful after a good grooming.

I was flattered to find that she was becoming attached to me. She seemed to like human company and I suppose, in the absence of other llamas, we were her herd, but in particular she followed me. When I went up to Carneddi in the mornings to see my mother and Mary, she would come too. She would wait for me in the field, no matter how long I was. Now there was no need to shut her in the old vegetable garden; I knew she wouldn't wander. There was no need for the collar and lead either. She would follow me, if she could, wherever I went, pacing along ten yards or so behind.

'She thinks you are her llamama,' said Paul.

I was pleased; it seemed a fine thing to be mama to a llama.

With each week that went by, we felt that Ñusta was more firmly established in health and that our experience of llama management was widening. Quite suddenly she discovered that grass was edible and began to graze. Certainly there wasn't much grass at that time of year but we felt that every addition to her diet would do her good. She was still thin but at least she was eating. Each small milestone brought us interest and satisfaction.

Another milestone was reached at Easter. We discovered that llamas like Easter eggs—or anyway ours did. Miles and J. came to camp with us for the Easter holidays as was now their wont, bringing with them large chocolate eggs for the children. These, with other eggs from well-wishers, provided them with a marvellous feast. When Ann and John had eaten as much as they could, there were still large pieces of egg in fancy foil tucked away in various corners. On Easter Monday the children were outside playing. Ñusta had come in for her swedes and carrots and had then retired to the sitting-room. I was washing up in the kitchen. While doing this, I heard various rustlings and shufflings coming from the sitting-room and went to investigate. I found the llama rummaging about on the Welsh dresser. She turned guiltily to look at me, and there was a piece of foil sticking out from the side of her mouth.

'Hey, you mustn't eat that! It's bad for you.'

I managed to grab the foil before it disappeared. Her nose went back to the remains of a shattered egg and she began to lip up pieces of chocolate rapidly. I managed to rescue the last of the egg—more worried that she would eat the foil than that she was eating the chocolate—and placed it on a higher shelf. Then it was a race round the room between the llama and me to see who could play 'hunt the Easter egg' the fastest. I won. When all the eggs were safely out of reach—quite a high situation was needed, I found, when dealing with a llama—I gave her a piece of chocolate on my hand. She guzzled it up with an enthusiasm I had not seen her show for food before. Here, obviously, was something that llamas really liked. The knowledge might be very useful when we needed to tempt or reward her at some time in the future.

Ann and John were pleased when they heard about the llama's raid on their eggs and hastened to give her some more. They thought it quite right that she should join in the Easter festivities—and I think they were satiated anyway.

# 9 Mooey

The llama now occupied a rather special place on the farm. She was an expensive, exotic and unusual animal. In buying her we had departed completely from all the traditions of the district. She represented a class of livestock about which we knew nothing and there were no local experts to whom we could refer, as we had always done with our problems in the past. There was not even a Ministry of Agriculture expert on the subject.

When my parents and I, backed up by the faithful Fred, had begun hill farming with no previous experience, we had attempted to follow the local farming traditions as closely as we could. Because of this, we found helpers and advisers on every side, and this is probably the reason why we managed to survive where sometimes others had failed. We had great respect for the local sheep farmers and followed their example as best we could, though we felt that we should never be as good at shepherding, even after decades, as those who had been born to it. The llama, however, was to be an addition to the traditional system farming, not the beginnings of a departure from it.

There were no wise Welsh experts on this branch of farming so we did not broadcast our intentions—or the llama's presence when she arrived. We didn't keep it secret either. We maintained at first what the newspapers call a 'low profile'. If llama-keeping were to prove a folly, it was better for us to realize the fact before anyone else did. If it were to be a success, time was needed to prove this. Of course we believed that it would be a success. We felt there was every reason to suppose that Ñusta would be a useful farm animal, giving us valuable wool, offspring and service as a pack carrier. Perhaps even more important than this was that keeping a llama was fun. I don't think anyone is in hill farming for the money; they do it because they like it. As old John Williams said to me long ago: 'I don't look to make money out of the sheep. I keep them for the pleasure.' Llama-keeping was a pleasure too.

I had hoped that our neighbour, Jones Williams, a clever and experienced shepherd, would one day come unexpectedly on the llama in the field. At a first quick glance, from a distance, she looked rather like a sheep. At that time her body was the same size and, in those surroundings, it was a sheep that one expected to see. At second glance, the long neck, long ears and long legs showed that she was a sheep with a difference. I had hoped that I might have the opportunity to complain to Jones that one of his rams had got at one of our pony mares and *that* was the result. What was he going to do about it? But I never got the chance to try that joke on him.

When Ann and John told their school mates that they had a pet llama which came into the house, their friends quite frankly did not believe them. A pet lamb, yes—a pet llama, no. Some of the children could not imagine what manner of beast a llama was. In the face of their disbelief, Ann and John had difficulty in describing it to them. This made me think of writing to the producers of the children's television programme, *Blue Peter*, on which unusual pets were sometimes shown, suggesting that viewers might like to see pictures of a really domesticated llama in Wales.

A week passed, two weeks, and there was no reply. Probably llamas did not fit into the current series of programmes, I thought. Then the phone rang. It was Mrs Evans, the Post Office, to dictate a telegram. 'TELEPHONE BLUE PETER SOONEST' said the telegram and gave the number. In fact lunch was ready so I thought 'soonest' must mean after we had eaten. When we had, I telephoned. The producer was enthusiastic about our llama. Could we bring it to Shepherds Bush shortly to appear on TV live? I said I didn't think we could. This request rather took me by surprise. I had imagined that a camera crew might be sent to film the animal in its Welsh hillside home, which would certainly provide a lovely setting. But I learnt that this would be too expensive; the television budget was tight. Camera crews were not sent out for anything less than three minutes of film. A llama probably was good for only one minute of film so we must bring it to the studio. Travelling expenses would be paid though there was no payment for the appearance. Would I think it over and ring again? I said I would.

Paul and I discussed the proposition. From any angle it didn't look very attractive. Of course it would be immense fun for Ann and John, but there might be other opportunities later. At the moment the elderly Land Rover needed a lot of repairs which would have to be done before it could take us to London reliably. The trailer wasn't in good condition either. Perhaps the factor which influenced us most against the idea of going was that we had had the llama for such a short time. She was just beginning to settle down but she was still thin and frail. It seemed unfair to subject her to any more traumatic experiences unless they were essential.

The possibility of going to London by train, however, was worth exploring. When we had moved from Nottingham to Wales, we had brought our two nanny goats with us in the guard's van. They had travelled extremely well and I had been able to be with them most of the time. Apart from eating a few labels off the luggage, they had behaved nicely and seemed to be quite happy. The idea was worth investigating so I telephoned the station at Bangor.

'Could you tell me, please, if goats may travel by train?'

There was a faint chuckle down the line, but the clerk went to look up the regulations concerning goats.

'Yes, goats may travel in the guard's van,' he said.

'Well, actually, it's a young llama I want to take, but she's the same size as a goat, only a bit taller.'

'Llama? Did you say llama?'

'Yes, that's right. Llama. But it's only the same size as a goat.'

'No, that would be different. I'd better look up the regulations on wild animals. They have to be enclosed in crates.'

I didn't think Ñusta would like to be enclosed in a crate.

'Llamas are not wild animals,' I told the clerk. 'They have been domesticated for hundreds of years and are quite ordinary farm animals.'

I went on to give a brief history of the llama's domestication. The clerk seemed quite interested.

In the end he said: 'I'd better contact Head Office in London and I'll ring you back tomorrow.'

He did ring back the next day. Head Office was quite definite: llamas were not allowed to travel by train. So that was the end of it. I wrote to the Television Centre explaining our situation, but we didn't hear from them again. Paul and I were not particularly sorry. We had been briefly in the limelight when my previous two books were published and we had found that publicity was a very mixed blessing. In a way it was fun but you generally ended up by being mis-reported or out of pocket, or both, and a lot of valuable time was wasted. Even so, it interested us to take a glimpse into so different a world from ours from time to time.

As I have already said, Ñusta was a special animal on the farm, but some of the others were too and it was not because they were exotic or unusual, but because of their characters or circumstances. Mooey, Ann's tame sheep, was one of these. We were amused to witness the first meeting between Mooey and the llama. For some reason their paths had not crossed until Ñusta had been with us for two or three weeks. At that time, Mooey was a sturdy six-year-old matron, rather more than middle-aged for a ewe. She would eat out of our hands and was not afraid of any dog. J. was here on holiday and he and I were walking up to Carneddi to see my mother. The llama was following us, twenty yards behind. I noticed Mooey by the garage, one of her favourite resting places.

'Hello, Mooey,' I said.

'Blair,' said Mooey.

She always answered if you spoke to her, and Ann and I could recognize her bleat amongst dozens of others. J. and I stopped to talk to Paul who was working in the carriage shed. The llama came walking past us, her attention fixed on Mooey. She still had a half-hopeful belief that sheep might, in the end, turn out to be other llamas, but they invariably ran away before she could investigate them thoroughly. Mooey, however, stood her ground. She was not an animal to be easily intimidated. The llama approached very cautiously, slowly lowering her head as she came near, till it was on a level with Mooey's own. I was aware that J. was getting his camera ready. The noses of the two animals touched. Click went the camera. Ñusta withdrew very slightly but continued to stare into Mooey's face intently for a few moments. Mooey still stood her ground. Then

Ñusta turned away, evidently satisfied that this animal was not another llama.

'I'm glad you got that picture,' I said to J. 'It's nice to have one of the llama and the old Moo together.'

We were all very fond of Mooey. Unlike some of the other pet lambs which we had reared to maturity, she was not much trouble. She would stay up the mountain with the other sheep in the summer and, during the winter, she did not break into the garden. The dairy nuts and hen food which she managed to steal were amply paid for by the fine twins that she reared most years. Also, she reminded me happily of the days when Ann was a tiny girl; the two had grown up together.

Mooey was born in the spring of 1969, when Ann was three years old. We often had one or more pet lambs in the springtime but usually managed to foster them after a few days to a sheep which had lost her own lamb. Pet lambs are lovable but they are also time- and milk-consuming creatures; they besiege the front door, dash in as soon as it is opened and generally manage to leave a trail of pellets and a puddle before they can be got out again. On the whole, we prefer not to have pet lambs.

I heard this particular lamb crying on the other side of the box hedge outside the kitchen window one evening in early April that year. It was time to feed John, who was then only a few months old, and Ann's suppertime too, but I hurried out of the cottage to have a quick look at the lamb. It was in the grassy yard at the back, though it scuttled away when I approached and wriggled under the gate to the Tŷ Mawr fields. Before it disappeared round the corner of the wall, I just had time to notice that it was fairly recently born, was strong and lively and that it had a black spot the size of a shilling (or perhaps I should say 5p) midway between its left eye and ear. I should know it again easily. There was no sign of its mother nearby. By now I could hear hungry wails from the cottage so I went back indoors. When the children were fed, bathed and tucked up in bed, Paul had come in from milking and was ready for his supper. In the hurry of domestic tasks, I forgot all about the lamb.

The next morning, when cooking breakfast, I again heard faint bleats through the kitchen window. A quick look outside showed that the same lamb was back again in the yard. There was no ewe which was obviously its mother in sight. Later in the morning, when John was having his sleep, Ann and I set out to try to find the mother. It was wet and muddy and a bleak day with a cold wind blowing. We proceeded round the nearer Tŷ Mawr fields but did not like to go out of earshot of the cottage in case John woke. Ann was in danger of losing her little gum boots and I had one ear cocked for sounds from the baby. It was not an ideal way to go shepherding. We didn't see anything much. Some of the sheep had lambs, some were yet to lamb.

There was a young-looking ewe not far away. Perhaps that was the mother. Sometimes a young ewe would lamb with little or no milk and afterwards take no interest in its offspring. We tried to drive the lost lamb towards this sheep but it would not co-operate and wanted to follow us instead. The ewe ran off and disappeared over the hill. I really ought to fetch Ruff, my old sheepdog, get some of the sheep together and make a serious effort to find the mother, but it was difficult to do anything quickly with a small child stumping along beside me. A few hundred yards of boggy, tussocky ground was quite an obstacle for her and could only be negotiated slowly.

'Can't *we* feed the lamb?' said Ann. 'Its mummy doesn't want it.' 'Perhaps it's a twin,' said Ann, 'and its mummy doesn't want two babies.' 'I should like a lamb,' said Ann. 'I would feed it. I haven't got a lamb.'

Last year she had had one and already she knew the delights of little lambs, but that one had grown big and now no longer needed a bottle.

'I feel cold,' said Ann. 'I've lost my wellie,' said Ann, 'and my sock's wet. I want to go home.'

'All right,' I said. 'We'll take the lamb and go back home.'

I retrieved the wellie, caught the lamb without much difficulty and we went back to the cottage. Ann stopped feeling cold and forgot about her wet sock. I prepared two bottles, one for John and one for the lamb. I explained that John's bottle was for John and the lamb's bottle was for the lamb and we must never mix them up. The lamb finished its bottle first.

It was probably a deserted twin and, though I kept a look-out for a ewe which might be its mother, I never saw one. Ann was very pleased to have a new lamb.

'What shall we call it?' I asked her.

'Mooey,' said Ann without hesitation.

She never gave a reason for her choice of name.

Mooey turned out to be female. At night she slept in an old tea-chest in the barn. During the day she spent most of the time under my feet. Ann fed her four times a day and sprinkled milk liberally on the floor too. I gave the late-night feed.

Ann loved the lamb. She found that if she bounced down the green slope in front of the cottage, Mooey would come bouncing after her. Together they bounced among the yellow daffodils of spring. Where Ann went, the lamb followed.

Later in the spring we had a litter of four puppies. As soon as they could walk, they followed Ann too. There was a big outcrop of rock at the bottom of the field in front of the cottage. On the far side of it there was a cliff but, on the field side, part of it sloped smoothly down to make a natural slide, seven feet long. Here Ann, the lamb and the puppies played together, climbing up and sliding down, sometimes on purpose, sometimes by accident. I explained to Ann that she mustn't go near the far side of the rock.

'I shan't fall down there,' she said, 'because I'm a big girl.'

She didn't fall, but she wore out several pairs of dungarees, scraped her shoes and polished the rock to a slippery smoothness. Mooey and the puppies grew big and Ann grew more adventurous.

In October it was time to send the season's crop of ewe lambs away for wintering. It is the practice on hill farms to send the lambs away to a lowland farm from the end of October till early April. Here, there is no competition with adult sheep; they are away from the rams and can grow well without the burden of pregnancy. Good wintering each year for the ewe lambs makes for a better-quality flock.

Ann was quite happy for her friend, Mooey, to go away with the others for their winter holiday. By now Mooey was a well-grown lamb with a good fleece. She was off the bottle but had developed a liking for dairy nuts, and she regarded herself more as a little girl than as a sheep. We had some difficulty in getting her into the trailer along with the other lambs. Ann came with us to the wintering farm to oversee the proceedings.

We unloaded the lambs in their new field and, at once, they began to graze the fresh grass—all except Mooey. She looked as though she didn't want to stay. We shut the field gate hastily, got into the Land Rover and drove away. The field bordered on the road and, as we drove off, we saw Mooey racing after us on the other side of the hedge, bleating loudly.

'Serves you right,' said Ann heartlessly.

Early in the New Year we went to the wintering farm to dose the lambs and see how they were getting on. Mooey had grown. She seemed to have settled down but was still rather separate from the flock. She didn't appear to bear us any ill-will and accepted a few nuts from Ann. Then we went away again.

At the appointed time in April, we fetched home the wintering sheep, now grown to the status of shearlings. We dipped them and put them up the mountain, all except Mooey. Ann wanted to have her old playmate near the cottage again, so we turned Mooey through the lower gate. By nightfall she had gravitated down to her old haunts, our front field, the sliding rock and the group of gnarled plum trees near the wall. Ann's old companion now bore no resemblance to the little white lamb that had skipped with her among the daffodils the year before. Ann herself was taller but Mooey seemed huge. She had grown into a strong shearling with a big fleece and a mature-looking gingery-coloured ewe's head. The black shilling mark was still visible between her left eye and ear. Ann was delighted to have her home again and fed her with dairy nuts. She went to bed saying: 'I'm so pleased to have Mooey back.'

In the morning it was raining. Ann looked out at the sodden field.

'I wonder where Mooey is,' she said.

There were a few ewes to be seen but Mooey did not appear to be among them.

'I expect she's sheltering somewhere,' I told Ann.

The next day was finer. Ann went to school in Beddgelert in the mornings now but I fetched her home at noon each day. Half a day at school, after the quiet of Tŷ Mawr, seemed to be enough for her at the age of four. In the afternoon Mooey still had not reappeared. I felt puzzled and a little worried. After dipping, she had come down to the cottage of her own accord and had seemed all ready to take up her life there again. Where could she be?

Two or three more days passed, and still there was no sign of Mooey. Now I felt really worried. I asked Paul and the farm pupil to look out for her but they could not see her. Perhaps she had found her way into the forest below our bottom fields and stayed there because there was plenty to eat. Perhaps she had gone up the mountain to join the other shearlings though this, I felt, would be out of character. Perhaps she was caught in brambles and unable to come back to us. This, I thought, was the most likely reason. Shearlings are athletic and they have a great deal of wool and not much sense. They are usually the ones to get entangled in brambles or wire. I told my fears to Paul and asked him to look again in all the most likely places. I was longing to go myself but it was difficult to find an opportunity. John was now eighteen months old, an unsteady walker but a hefty child to carry. An expedition with him was hard work and I never seemed able to cover much ground. The torrential rain made matters worse.

Paul looked again without success and Ann kept asking: 'Where's Mooey?'

About a fortnight after Mooey's disappearance, the weather improved. The wind dropped and evening sunshine lighted the fields and bare woods, where the larch trees were just beginning to show a hint of green. John had his supper early that evening, and was bathed and asleep in bed by six o'clock. It was still a lovely evening.

'Can we go for a walk before bath-time?' said Ann.

'Just a short one, if you like,' I said. 'We'll go down to the Gors and get a log for the fire tonight, but we mustn't leave John for long.'

We put on our gum boots and went down through the gap in the wall, past my old Writing Hut, and down over the rocky shelf which dropped into the Gors. Gors Goch, the Red Marsh, ran along the lower boundary of our land, a narrow valley of peat, grown with rushes and tall tough grass. It was a secret and unfrequented place, bounded on the upper side by precipitous oak woods and on the lower by the forest. Here there were good branches, blown down by the gales, that could be gathered up for firewood. I found a suitable log and Ann began to tow a large branch behind her.

'These will be nice for the fire,' I said. 'Let's go back now.'

'Let's go the long way round,' said Ann.

The long way round meant walking along two-thirds of the length of Gors Goch, passing the old ruined shed which had once housed cattle and hay in bygone days when the field had been mown with scythes. It meant climbing up a steep and stony track and traversing along the top of the oak woods to the cottage. It would take us quite a time at Ann's rate of progress and encumbered by our firewood.

'I think we ought to go back to John.'

'John's asleep and doesn't know we've gone,' said Ann. 'Please let's.'

She was quite right. When John was put to bed, he went to sleep and didn't wake up till morning. It was a lovely evening after all the rain, with a smell of spring in the air. Though most of the branches were bare and the colours of autumn were still present in last year's dead bracken and fallen beech leaves, the larches were turning green and the buds on the birch trees had the purplish sheen which precedes the appearance of tiny leaves. It was a beautiful evening for a walk.

'Well, all right,' I said. 'But we must be quick.'

We set off along the hidden valley with our firewood. When we were about halfway along it, Ann said suddenly: 'There's Mooey.'

I looked where she pointed and saw a grey sheep lying down just before the beginning of the stony track by the ruined shed. There was a white object close to the sheep.

'It can't be Mooey,' I said. 'It's got a lamb.'

'It *is* Mooey,' said Ann.

We hurried eagerly forward to look and indeed it was Mooey. There was the black shilling mark and the lamb was hers too, though how she had got in lamb when she was away at wintering we were never to know. But there was something wrong. Mooey did not get up to greet us, though the lamb ran away for a few yards. It was the tiniest lamb. Mooey's voice was only the faintest whisper when she said: 'Blair.' Then I saw what was wrong. A short length of blackberry briar was entangled in the long wool of her neck and throat. It was twisted round and round and round into the wool to make an unbreakable bond which held Mooey prisoner. It was clear that she had struggled long and hard to escape. The full seriousness of the situation became apparent when I noticed that the ground, in a small circle round the briar, was nothing but bare and churned-up earth. There was not a blade of grass or a leaf that Mooey could reach. She must have been there for days, circling endlessly and winding her bond forever tighter. How Paul and the pupil could have missed her, I could not imagine.

'Poor Mooey!' I said.

'Poor Mooey!' said Ann. 'We'll rescue you and give you lots of food.'

I began to hack at the bramble with my pocket knife. It was a terrible job with so many briars wound deeply into the wool, but at last she was free. She was too weak to stand. She must have lambed while a prisoner, for the lamb was very young. Another day or so and they would both have been food for the crows.

'We must take Mooey back to our house so that we can look after her,' I told Ann, 'but we must be quick because of John. I think I can carry her if you can manage the lamb. You mustn't let it go because we might lose it.'

Cutting the bond must have taken some time, for already the sun had dipped behind the mountains and the light was going out of the day. Then, as later, Ann was very cool and effective in an emergency. I put the lamb in her arms. It was only a tiny creature but it wriggled violently. Ann was small too but she held on firmly. Then I picked up Mooey. She was a good-sized yearling with a huge fleece but, under all the wool, there seemed to be only bones. She didn't weigh much and she didn't struggle. We set off up the steep rocky steps of the path, the firewood abandoned. Mooey seemed to grow heavier as we climbed up and I couldn't see where I was putting my feet. It was hard going.

'Are you all right, Ann?'

'Yes.'

We hadn't much breath to spare. The way seemed endless but at last we reached the cottage and put our burdens together on the grass outside.

'We've saved Mooey, haven't we, Mummy?' said Ann.

'Yes,' I said and hoped that my words were true.

I had a quick look upstairs and John was still sleeping peacefully. Then we set about restoring Mooey. I stood her on her feet. She balanced there and then began to graze. Ann rushed off and returned with a generous supply of dairy nuts. Mooey began to eat them ravenously but, after only two or three mouthfuls, she turned away and would eat no more.

'Come on,' said Ann, 'you like those.' But Mooey would not be tempted. We offered her a drink but she didn't want that either. She pottered off slowly down the field with the lamb following.

'Let's leave her to do what she wants,' I said. 'It's your bath-time now.'

Ann went to bed radiantly happy.

'Wasn't it a good thing that we went the long way round?' she kept saying. 'We might never have found poor Mooey but now we've saved her.'

It was indeed a very good thing that we went the long way round but I wished we had gone that way sooner.

Ann went off to school the next morning and later I found Mooey sitting under a tree with the lamb beside her and a glassy, sunken look in her eyes. It was going to rain soon so, very slowly, I coaxed and chivvied the pair of them up the slope and into the stable.

It is difficult to reverse the effects of extreme starvation and I knew that I should have to use every trick I had if Mooey were to live. In those days, I didn't know about the almost magical effect of a vitamin $B_{12}$ injection in such cases. If I had, the task would have been easier.

I didn't see how Mooey could feed the lamb in her present condition but I felt she should have it with her as an incentive to live. It was a lively little creature, although so small and frail. The pupil named it Cuthbert. I tried to give him a bottle but he was one of those lambs which violently reject a rubber teat. I didn't persist too long. I felt that, at that stage, too much coughing, spluttering and wheezing would do more harm than good and might result in a case of pneumonia. Then I offered Mooey a variety of her favourite foods but she wouldn't eat. The only hope was to administer my special potion which had been fairly successful in the past. This was a mixture of warm water, milk, egg white, glucose and glycerine, given as a dose in small quantities at frequent intervals. Good nursing would be her only hope.

So, for the next three days, I visited Mooey with the potion, at three- or four-hourly intervals, turned her on her bed and talked to her. 'You must get better. You must survive after all this.' I couldn't forget Ann's joy at finding her old friend again nor her determination in carrying the lamb safely home.

The bottle-feeding of Cuthbert was not very successful. He choked, gasped and showed every sign of loathing my attempts to feed him. He preferred to suck at Mooey's seemingly empty udder as she lay there—she was too weak to stand now. He didn't grow but he didn't seem to become any more feeble, so I left him to it.

After school, Ann visited Mooey each day in the stable. She was concerned by her sheep's illness, but did not seem to doubt for a moment that Mooey would recover and that it was only a matter of time before they would be playing together on the sliding rock.

On the third evening, I squatted by Mooey in the bracken, giving her the last dose before going to bed.

'You really must get better for Ann's sake,' I told her.

While there's life, there's hope, I thought, but I had had too many cases of sheep that had died to be complacent. With a sheep like Mooey, the human contact meant something and, as I crouched there, I willed her to live. It was very quiet. A little mouse popped its head out of a chink in the stone wall nearby, looked at us and disappeared again. The light from the electric bulb did not quite illuminate all the dark corners of the ancient stable. The cobwebs in the roof palpitated slightly in the draught from the open door. Was it imagination that made me think there was a little more light in Mooey's golden eye and that she was showing a feeble enjoyment of the liquid that I was trickling into her mouth? I couldn't really tell, but I went back to the cottage feeling more hopeful.

The next morning the signs of improvement were marked. Mooey raised her head when I entered the stable and accepted the drenching bottle into the side of her mouth with something like eagerness. This was what I had been looking for.

'You good old sheep,' I told her. 'I think you're going to make the grade.'

She even managed a little maternal bleat to Cuthbert. When I fetched Ann from school that day, I was able to tell her that Mooey was much better.

It was a slow process. The next day Mooey was able to get on to her feet and to eat a little hay and ivy. She still wouldn't eat concentrated food. The day after, we let her out to graze because too much sloppy and unnatural food is bad for a ruminant digestion. She was feeble and weak but she managed to potter along. The weather had improved, with milder nights and sunshine during the day. We decided to leave her out that night so that she could have more grazing time. The next morning Mooey had gone again.

Paul was now thoroughly alerted to the situation and hurried off to find her before another tragedy could occur. He discovered her and the lamb and drove them slowly home. She had retreated to a rather odd place. There was a deep, tree-shaded ravine on the lowest easterly boundary of Tŷ Mawr land. A stream trickled down the middle of it. There was not much grass or bracken under the summer canopy of leaves; and the ground, the fallen branches and the boulders were covered mainly with moss and lichen. It was a gloomy, lonely place, where the sun did not shine and the trees were very old. Here Paul found Mooey and her lamb among the rocks and tree trunks. She was a long way from the cottage and there was little there for her to eat, so he brought her home. She went off again to the same place the next day and the pupil brought her home. Ann and I fetched her the third time. After that, when she again returned to the ravine, we left her there. There must have been something about that lonely spot that she needed for her recovery; it had never been a favourite haunt of hers in the past.

Each day, after John was asleep in his bed, Ann and I would take our evening walk to Mooey's retreat. I should have preferred to go alone. Ann was upset by Mooey's departure but each day she insisted that she saw her pet before nightfall. I was afraid of what we might find and, on each visit, it was a relief to come on Mooey's sad grey figure between the tree trunks with the small white Cuthbert by her side, and to know that they had both survived.

'Mooey!' Ann would cry, and 'Blair,' answered Mooey in a feeble voice.

The brussels sprouts in the garden were now running to seed. We would take a massive stump, with its load of greenery, as an offering for Mooey, and this she would eat. She also began to take a few nuts from our hands. We tried to lure her back towards the cottage. She would follow us for a few yards and then stop. No amount of coaxing would tempt her any further.

'Mooey, Mooey,' Ann would call. 'Come on, Mooey.'

'Blair,' said Mooey in her weak voice.

'Leave her there,' I said. 'She wants to be there to get quite better. It's nice and quiet for her.'

As we looked over our shoulders, we could see Mooey standing under the trees, looking after us but not attempting to follow.

'Good-bye, Mooey,' called Ann.

'Blair,' said Mooey mournfully.

'Good-bye, dear Mooey.'

'Blair.'

And so it went on, until Mooey was lost to sight among the trees, until her sad bleat was inaudible and tears were flowing down Ann's cheeks. It was quite an upsetting experience to take the brussels sprout plant and the handful of nuts each evening.

After a few days, we found that Mooey was following us a little further each evening. At last the day came when we managed to coax her right out of the ravine and on to the open lower land. The next night she came as far as the bottom end of the Weirglodd, the flat field to the south of the cottage. Before Ann got into bed that night, I lifted her up to look out of the bedroom skylight and there, at the far end of the field, she could see the grey blob that was Mooey and the white dot that was Cuthbert. They had almost come home. Ann went to bed happy that night, the first time for many days.

After this Mooey improved steadily and was her normal self again by the summer. The rescue was quite complete.

# 10 Llama at Home

We had brought our llama from Knaresborough to Wales in January 1975. In the following April my sister, Mary, died. The miracle of her recovery, which we had longed and prayed for, did not happen. The beginnings of her brilliant career never came fully to fruition through her illness. She was only four years older than I. The last five or six years of her life had been so miserable that no one could have wished it to be prolonged, yet always we had hoped for a wonderful recovery. My mother had worked for it tirelessly and courageously through long days and nights. It was bitter for her to be bereaved twice in two years, and perhaps the bitterest bereavement any woman can suffer is to lose her child. As I looked at our own two children, I realized it was a bereavement from which a woman could never quite recover, however much time passed. Now I was all that my mother had left and I was very glad that I had not died at the age of seventeen, as I so nearly had all those years ago. She needed me now as she never had before. Fortunately we had always been very close and we were able to prop each other up in time of need. Paul, as always in a crisis, was a powerful support. So was Beenie.

Now my mother demonstrated to us that life must go on whatever sorrows there were. She set to work to rehabilitate her garden which had long been neglected. She had more time for her grandchildren. She took a renewed interest in the farm and pleasure in the llama. We all drew inspiration from her courage and unbreakable spirit.

After the children were in bed, I would go up to Carneddi as usual to spend the rest of the evening with her. Now the llama would often come too, and Mother liked me to invite her indoors. Ñusta was really living up to my hopes for her as a new interest for us all, a fun animal, a refreshing departure from tradition and a step out of old ruts. She would bustle into the Carneddi kitchen behind me and pause to inspect my mother. She had an odd little manner of looking into the faces of the people she knew, especially if they were seated. She would curve her swan neck so that her face was directly in front of theirs and, for a long minute, would gaze into their eyes with her ears forward and her nose a fraction of an inch from theirs. She remained perfectly motionless while this communion was taking place. Sometimes, when we were all sitting at the table at Tŷ Mawr, she would go from person to person, communing with them in this way. I felt she was trying to tell us something, but what it was I did not know. She seldom did this with strangers.

She was not so much at ease in the Carneddi kitchen as she was at Tŷ Mawr, because she was less used to it, but she liked coming in all the same. She would tour the room, inspecting everything. She took a fitful interest in the tele-vision set, when it was switched on, and seemed to listen to and like classical music, though pop made her flatten her ears. Paul and I had decided against television at Tŷ Mawr. We felt it was a time and money waster, a brain-washer for the children and an interrupter of our compulsive reading habits. With a set at Carneddi, however, we had the best of both worlds. Ann and John could see enough children's television to keep in touch with their contemporaries without becoming square-eyed, and the adults could arrange to be there to see any specially interesting programmes. My father had a set installed as soon as we had electricity, just before Ann was born. After twenty years of living with oil lamps, he had thoroughly enjoyed the conveniences of electricity and he found television among the best of them. It gave him much interest and entertainment in his last years.

I was sorry that he had not lived to see the llama. After a short period of being anxious at the prospect of having so strange an animal and doubting the wisdom of it, he would have fallen under her spell, for he was a great animal-lover. He would have begun to dote on her. He would have bought sweeties for her and had private communions with her while she ate them. He would have referred to her as 'my llama' while astonishing his acquaintances with some of her exploits.

It was a comfort that Mary had seen Ñusta. She loved animals too and particularly liked elegant and fluffy ones. The llama was just Mary's kind of animal and she described her as 'heavenly'.

In the evenings, when Ñusta had examined the kitchen and eaten any offerings of apple peelings which Mother had left for her, she would fold up and sit down with a slight thud on the red-tiled floor. She generally chose a position facing the door and close to where Mum and I sat on the settee. From it, she could easily nibble Mother's tweed skirt, and this was a favourite pastime. She had this odd habit of liking woolly materials, which she would nibble and pluck with a quick, gentle and continuous movement of her prehensile lips. She would pluck and eat all the little bobbles that form on woolly material. It was a sort of grazing and she did a good tidying-up job on several of my older sweaters. The quick tickling movement of her lips was quite pleasant—one felt a little honoured to be given this attention—and she did not damage the material or even dampen it. She liked Acrilan jumpers too. Paul said that, if she ate much of it, her own fleece would turn into Acrillama.

It was at Carneddi that we discovered that Ñusta also liked newspaper. There was a lower shelf to the table near the settee and here Mother kept old newspapers, outdated magazines and past copies of the *Radio Times*. From her sitting position near our feet, the llama could easily stretch out her neck and pull the piles of papers on to the floor, rip off a few pages and munch them up. My goats had liked eating paper but never in the quantity that Ñusta seemed to want. A whole *Daily Mail* or half a *Radio Times* was quickly gone. At first I was worried that such a quantity of paper might harm her or that the newsprint was poisonous, but she ate it regularly with no ill effects. In fact she was avid for a certain amount of paper every day and I came to the conclusion that she must need it. Later I asked the vet about the edibility of newspaper and was told that experiments in feeding pulped newspaper to cattle were being carried out. Apparently the cows did quite well on it. Then I read that llamas were able to utilize crude fibre at least twenty-five per cent more efficiently than sheep. There must be lots of crude fibre in newspapers, I thought, so she must be getting some nourishment from them.

This liking for paper could have awkward side effects. It was all very well to explain to a friend that, sorry, you couldn't answer his letter properly; the llama had eaten it. The explanation might be less acceptable, I felt, to officials or senders of bills. They might regard it as a tall story. It was very tantalizing to catch Ñusta with the last corner of a piece of paper disappearing into her mouth. What communication was it that had gone for ever? There was no means of finding out. Paul had the idea that we might popularize llamas in business offices instead of waste-paper shredding machines. We became very conscientious about putting our important papers away, though this was not so easy as the llama grew taller, in our small and overcrowded cottage.

We partially solved the problem by giving Ñusta her own supply. Mother's shelf for old newspapers kept her happy at

Carneddi and, at Tŷ Mawr, I provided her with a toy-box. This was a large cardboard box with a pile of magazines in it and sometimes a heavy log or two. She liked to nibble the logs and their weight helped to hold the papers so that she could rip them up. She liked the cardboard box best of all. She would tear strips off the edge of it and in two or three days she would need a new one.

After a while she seemed to connect the words 'toy-box' with the object. If we wanted to divert her from the valuables on Paul's desk and suggested that she should play with the toy-box, more often than not she would hesitate as though thinking and then turn round and trundle over to it. When she was in the mood, she would settle down on the mat for a good play. It wouldn't be long before the logs were tipped out and the floor strewn with magazines and half-eaten pages. With the weight of the logs gone, it was harder for Ñusta to tear bits off the cardboard box. She managed by biting the edge of it and shaking it about, as a terrier shakes a rat. It was strange to see her sitting there, whirling the box round her head. Once or twice it dropped completely over her head and, for a moment, she sat, quite extinguished, before tipping the box off again. She liked John's plastic lorry too and, if it were near her on the floor, she would take hold of it and zoom it backwards and forwards.

On my nightly visits to Carneddi, I noted that Ñusta was afraid of the dark. The evenings were getting lighter and the sun had hardly set on our way up, but when we came down again, it was often pitch dark. She followed me everywhere and liked to know where I was. If she mislaid me, she would hurry round looking and tooting with anxiety. Her normally low-pitched and musical *mmm* would turn into a desperate little shriek on a rising scale. But if she knew where I was, even if she couldn't see me, she would settle down to wait for an indefinite period. At first I had taken her into the kitchen with me but later in the season, when Mother had pot plants and seedlings ranged round the room, I left her outside. She seemed quite happy, sitting by the garden gate under the shelter of the fir trees, but I noticed that, at night, she never sat in the open. She sat with her back to the wall and kept a constant look-out. Had she, I wondered, an ancestral fear of predators or was she less secure because she was not part of a herd with a protective male at its head? We sometimes wondered if the thought of pumas caused this ancestral fear because she showed a very marked interest in cats. The interest did not seem tinged with fear but cats obviously meant something special to her. She would watch their every movement with close concentration. Of course it may just have been that she was remembering her near neighbour at Knaresborough, the lion.

Her interest in cats was in contrast to her indifference to dogs. These she ignored unless they got in her way. She was a little wary of the cattle and ponies because they were large and sometimes chased her, though she liked Ann's pony,

Dolmen. He was a gentle animal and the two would touch noses and commune with each other.

When I came out of Carneddi at night and through the garden gate, I would hear a low toot and Ñusta would be by my side. Her normal walking pace was slower than mine but in the dark she would hustle along and keep close to me. I thought she didn't see well at night as she would sometimes hesitate over the rocks and get left behind. Then I shone my torch back and called encouragement. Unlike the eyes of dogs, cats and sheep which glow green, her eyes shone red in the beam of the torch. She would give an anxious toot, scramble over the obstacle and race to my side again. When she was older, she grew much more confident in the dark and did not seem to need my proximity so much.

One evening I was walking down from Carneddi in the gathering dusk. There was still light in the sky but all the colour had gone out of the landscape. The mountains were dark silhouettes and the distant lights of Harlech showed bright to the south. I looked back to see if the llama were following. I was descending the green path to Tŷ Mawr and I saw her just coming over the brow of the hill about thirty yards behind me. She was silhouetted against the sky. I never looked at her without thinking that she was strange and beautiful, but now she looked strange indeed. She was coming directly towards me but the light was so poor that it was impossible to see her forward progress. For a few long moments she was just a tall silhouette which seemed to undulate. The long ears, looking like a short pair of forked horns, were outlined against the sky. The pale figure seemed to glimmer, seemed for a moment semi-human with the dark splodges of eyes, guessed at rather than seen. She looked like some faint wraith. If I had not known all the time that she was my llama, I should have been frightened out of my wits. I hoped that some benighted walker never met her thus in the gloom.

In a few moments she had crossed the skyline, broken into a galumph and was at my side.

'Well, you are a peculiar animal,' I told her.

'*Mmm,*' said the llama.

Ñusta spat for the first time in May that year, when she was about eight months old. We had more or less forgotten about the spitting business and it took us by surprise. A Dutch friend was having tea with us. Brenda and Helen had come up from their house in Nantmor for the afternoon. The llama had walked in, been admired and had sat down on the mat. It was obvious that she intended to be part of the gathering, but after a while I could see that she was feeling rather agitated. The dogs kept coming and wagging round the visitors. Helen and John were playing noisy games on the floor nearby. The llama looked as though she felt threatened and hemmed in. She kept flattening her ears and raising her head up high. Any minute I expected her to get up and move to a more peaceful spot. Then Helen took a backward step, tripped and fell right across the llama's back. It was too much for Ñusta. Quick as lightning, she spat in the nearest face. Unfortunately this happened to belong to Els, our Dutch friend, who was sitting on the low couch near the llama. There was no 'ready, aim, fire' about it; Ñusta shot, as it were, from the hip without a moment's warning. Everyone was taken by surprise. For a second, Els didn't know quite what had happened. Her face and glasses were bedewed as from a fine green aerosol and an unpleasant smell, suggestive of rotting cabbages, filled the air. It was a regrettable and embarrassing incident, not the thing to foster international relations, I thought, but fortunately we had been friends with Els for a long time and I knew the schism would not be serious. Els had to wash her face and we apologized for the sudden and unladylike behaviour of our animal. I was very sorry that Els was the first victim but, underneath, I was pleased to know that our llama was capable of true llama-like behaviour. We had wondered what the spitting would be like; now we knew—sudden, insulting and smelly but not serious.

After this Ñusta spat when she thought the occasion demanded which, on the whole, was not very often. She would sometimes spit at the dogs if they crowded round her legs when she was eating from her bowl in the kitchen. Sometimes she would spit at people in similar circumstances. If she had a mouthful of nuts at the time, they would ricochet round the kitchen, like peas from a pea-shooter, in a most dramatic way. You really felt you were under fire. However, she generally gave warning of her intentions so that you had time to withdraw. Her ears would slam back and she would make snaking movements with her head, pushing her face into yours, munching, with her mouth slightly open and eyes goggling. Sometimes she would utter low grunts. The behaviour was so unlike that of any of our other animals, that it was easy to imagine she was trying to give you a quick kiss. If you didn't move fast, however, you would soon be disillusioned.

'A bit of spit and plenty of polish,' said Paul, wiping himself after one such episode.

It was always faces that she aimed at, but if you put up a hand or held a tray or similar shield between your face and hers, it would put her off her stroke. She needed to see the whites of your eyes before she pressed the trigger. This fact often let us get by unscathed. Some people found it convenient to bend low, avert their faces and scuttle past the danger area. It all added to the excitement of our lives.

Sometimes, when shut out of the cottage, she would spit at someone trying to enter. She often spat if she were caught raiding the sugar bowl or flower vase and someone rushed to save it. A quick, aggressive movement would usually trigger off a quick, aggressive spit, and this with no warning. You could never push her out of the way. A cross face would swivel round and might deliver a Parthian shot, to your discomfiture, but you could always lead her. If it was necessary to put her out, you could always loop a rope round her neck and, after a moment's hesitation, she would follow you meekly in the desired direction.

Sometimes she threatened me and sometimes she would spit at me. I was by no means exempt, but she did allow me to take more liberties with her than she allowed to anyone else. Sometimes, by just talking to her and not shielding my face, I could get her resentment to subside. She would stop snaking her head, her ears would come forward again and she would resume her eating quite happily.

The Director-Secretary of the Chester Zoo had referred to the 'nasty habit of spitting'. The spitting, we found, was nasty, but not very nasty. It was almost the only defence that Ñusta had and, without it, she might have seemed altogether too mild, too patient and too helpless. She might have seemed an animal of less consequence if she had not had this means of expressing her rich and varied emotions. It certainly added a spice of hazard to her presence in our home.

# 11 Coat and Slippers

Fred, my old nanny, and Norman, her husband, came over from Australia in the spring of 1975. It was Norman's first visit to the Old World, and we had not seen Fred since our wedding fifteen years earlier. Fifteen years is a long time but it seemed as nothing when Fred walked in through the front door. She was just the same old Fred who had presided over my babyhood, been interested in my school days, shared the hopes and fears of adolescence and helped to launch me as a hill farmer. She was a little greyer but the big dark eyes held their old fire. She was fatter too.

'Yes, I am fat,' she said crossly, patting her spare tyre.

But she was the same old Fred, ready to be teased and loved, and we began again just where we had left off. Norman was nice. He was just as I had imagined he would be from the photographs, the letters and the tape-recordings. The trouble with Norman, as far as I was concerned, was that he liked to go places and see things. This meant that he whirled Fred away to the farthest corners of the British Isles and Europe when I should have preferred her to be at Carneddi in the old pattern of our family life. It was selfish, I knew, and of course a sight of Fred was much better than nothing.

Fred was happy to be with her own dear lady, Mrs Ruck, after so many years, but she was sad that she had missed seeing my father. She did see Mary, but it was on the day after Fred arrived that Mary died. Then Fred was united with her old foster family in mourning. It was a great comfort to have her there at such a time.

Fred made a birthday cake for me that May, slipping back into the old ways of my childhood years. It was a good party, considering all that had happened. Brenda and Helen were there, and Norman, Fred and her sister, Elsie, and all our family. The cake was lovely, complete with the few candles that are obligatory, even if there isn't room for the correct number. Fred had made a little woolly llama to stand on the top, with a figure, representing me, offering a real Malteser to the llama. We all thought this was marvellous. Ann and John had not realized that Fred was so clever. For years I had been telling them stories about her and she had grown to be something of a legend. Now, having reached the ages of six and nine years, they were shy. They couldn't fall into her arms and hug her like I could. In spite of my stories, she was still a stranger to them at the moment but Fred had a great way with children.

Now Fred could catch up with all the news of the farm and see with her own eyes the developments that had taken place in the last fifteen years. She could see the ponies for the first time and particularly Princess, founder of our herd, now twenty-seven but looking years younger. She could see the new generations of sheepdogs—worse than the old, we sometimes grumbled—and meet Mooey, who reminded her of Topsy, our very first pet lamb. She could see the cattle, all descendants of the cows she had once known. We explained how we had repaired and refurbished Clogwyn, the remote little farmhouse further up the valley, and let it now as a holiday cottage. Of course the llama was what I particularly wanted her to see. By autumn, Fred would be 12,000 miles away in Australia, but at least she would have met our latest beauty.

The llama liked Fred and Fred liked the llama. Fred had always been good with animals. She identified with them and they responded to her direct and spontaneous approach. I let Ñusta into the Carneddi kitchen and Fred, Norman and Elsie were amazed to see such a large and elegant creature in the house. Ñusta inspected them one by one and then went on her usual tour of the room to see if there was anything good to eat.

All too soon Fred and Norman left Wales to continue their sightseeing elsewhere. It was sad to part but Fred promised that she would come again. We settled down to the farm work of late spring and summer, planting the small area of potatoes and kale which we grew each year, dipping the ewes and lambs and putting them up the mountain for their summer grazing.

At the end of June it was shearing time. We wondered if we should shear Ñusta along with the sheep but, in the end, decided not to. We were a little daunted by the difficulties of the task and we were not quite sure if it was the

right thing to do. Ñusta was not yet twelve months old, so
to shear her would be like shearing a lamb, which was not
our usual practice. One authority said that the Indians did
not shear their llamas but combed them to get the wool.
Another source said that they were sheared, just like sheep.
For the moment, we decided to leave the job. I brushed
Ñusta fairly frequently and saved the few brushings. Not
much wool came out but we thought that, at some future
time, she might have a definite moult and then we should
get more.

Beenie and I would have been reluctant to sacrifice the
beautiful fleece. It had grown a good deal in the past months
and now hung round the llama in elegant draperies. On a
windy day the long hair would ripple in the gusts like a field
of corn. Ñusta loved the wind and would romp and frolic
along, ears at a rakish angle and skirts fluttering round her.
Normally she moved in such a slow and stately manner that
it was unexpected to see this large animal suddenly galli-
vanting along in the teeth of the gale, like a huge autumn
leaf bowling before the wind.

Her long hair now reminded us of clothing. The white
of her neck finished in a perfect 'V' at her chest and the
greyish-fawn of her body met it symmetrically from each
side, giving the appearance of a shawl pinned neatly on
the bosom. Her legs were covered in short thick fur, like
whitish, mottled stockings, while the long hair of her body
hung in an even fringe above her knees, like a feathery

mini-skirt. For all her elegance, there was also something slightly comical about her attire.

Her whole body, and particularly her neck, was covered with long single hairs which blurred the outline and gave the appearance of a halo when seen against the light. Her tail was a thick mop of ginger hair. We called it the Feather Duster and she seemed able to do extraordinary things with it. When she batted it up and down, it looked very much like a duster being shaken from a window or a large hand waving good-bye. It seemed to be about twelve inches long, but she would not allow anyone to investigate its conformation closely. At times she would carry it over her back and then the bare skin of the underside was visible. Paul and John soon discovered that if this bare underside were touched, even very lightly, the feather duster would come slamming down into a shut position over her naked rear area in quite a dramatic way.

'Stop teasing the llama,' I would command.

Sometimes she wagged the tail briskly from side to side. It seemed to be a powerful and mobile member.

We had always referred to the llama's fluffy covering as 'wool' but now I read that it should properly be called 'fibre'. The same applied to the product of the alpaca, vicuna and camel. I learnt that, in fact, the fibre of the Camelidae resembled human hair rather than sheep's wool in molecular structure. Ñusta seemed to have two sorts of fibre in her fleece—long, fairly coarse, straight hairs with a mass of soft gossamer filaments between. Apart from the grass seeds and the odd bits of debris which it picked up and which I attempted to brush out each day, her fleece was amazingly soft and clean. There seemed to be no trace of grease in it, and your hand felt as clean after stroking her as it did before. She was a creature adapted for life in an arid climate so she had no need for grease to shed the rain.

The dryness of her fibre made us wonder how fitted a llama would be to the wet Welsh climate. Here we enjoyed ninety inches of rainfall a year and here it could rain for a fortnight without stopping. There was no doubt that Ñusta did not like the rain. She would galumph rapidly for shelter at the beginning of a downponr, with her ears held sideways and her head low. If the stable or cottage were not open for her entry, she would cramp herself into some small sheltered corner and remain there, with an expression of resignation, until the rain was over. Occasionally she had a real wetting and I would look closely at her fleece to see how she had fared. Although I supposed that it was not primarily designed to shed water, it seemed to do the job very well. The thousands of fine strands, massed together, collected the water like drops on a cobweb and the drops were led off downwards by the longer thick hairs. When I parted the fleece, the skin below was warm and dry. We found, however, that it took a long time for the outer layers of fibre to dry after the rain. Ñusta seldom shook herself.

She was well protected for short showers but I had the feeling that her fleece would not serve in a prolonged downpour. So we always put her in the stable on a wet night and left her in all day if it was pouring. The ideal would be to have a couple of open sheds where she could go if she needed shelter, one at Carneddi and one at Tŷ Mawr. These we didn't have but Paul said he would build llama shelters as soon as he had the chance. On a wet day, if the car were out, Ñusta would park herself in the garage with obvious satisfaction.

Another hazard for llamas in a wet climate seemed to be the condition of their feet. Mr Humphreys, the Colwyn Bay vet, had warned us that llamas might get skin diseases of the feet and lower legs in wet and mud. They did not, however, suffer from the same sort of foot-rot that could be painful and troublesome in sheep; the anatomy of a llama's foot was quite different. This, of course, we had noticed. Ñusta's strange feet were one of her fascinating features. They were surprisingly small for her size. Each foot was deeply divided to form two toes and each toe had a large, black, slightly curved toe-nail or claw. Her foot-marks looked rather like the slot of a deer except that there were two claw marks in front of each print. She certainly had not got hooves. The sole of each foot was covered in tough polished leather, very warm to the touch. Her feet, particularly the hind ones, looked ridiculously small on the end of the long, stick-like legs. They reminded me of the tiny bound feet of an old-style Chinese woman and they looked flat, as though she suffered from miniature fallen arches. I read that animals such as gazelles had tiny feet to reduce their inertia. Ñusta had not the agility of a gazelle and we wondered if this was true for llamas also. The design of her feet was such that, even on a very wet day, she left only dampish prints on the cottage floor, nothing like the muddy paddle marks left by the dogs.

On both sides of her hocks, Ñusta had long-shaped bare patches covered with a horny crust. These reminded me of a horse's chestnuts though they were obviously not the same. In a book, I found them described as 'glandular spaces in the metatarsal region', but the function of the glands was not mentioned.

We noticed that Ñusta's toe-nails were beginning to get long so Paul and I decided to trim them. We were reasonably expert at trimming the hooves of sheep, goats and ponies and, with some difficulty, I had even shortened the hooves of a cow, but a llama was different again. The hoof-clippers, which we used for the sheep, seemed likely to be the best tool for the job and more effective than a knife. They resembled a pair of secateurs with straight, pointed blades. With them, it was easy to snip off small pieces of horn. We put on Ñusta's collar and Paul held her while I tried to snip. Ñusta objected. She seemed to think we were trying to take liberties again and reared up on her hind legs. She was about seven feet tall when she did this. I hung on to a front foot and Paul managed to get her to go down again. Then she tried a rush forward, then a rapid reverse. Still we hung on. Finally she seemed to decide that she was losing dignity and that we weren't to be shaken off easily. She subsided to the ground with a very annoyed expression on her face. In this position, almost standing on my head, I managed to do the snipping, cutting off only a very modest amount of horn. The old advice of shepherds, to have a sharp knife and a hard heart when treating a case of foot-rot, was something I always disregarded. I didn't want to risk cutting the quick and, since I didn't know quite where it was, I erred on the safe side and cut only a little. However, when Ñusta stood up again, her toes looked very much neater.

While this operation was taking place, Beenie had been hovering round with a bag of Maltesers. The idea was that Ñusta wouldn't notice what we were doing to her while her mind was occupied by one of her favourite treats. It didn't work like that; Ñusta was not prepared to accept a bribe while she was suffering what she obviously considered to be an assault. But, when it was all over and she had risen to her feet again, she seemed ready to forget it all and accepted a few sweeties in a gracious and friendly manner.

As the weeks went by, we were pleased to see that the llama's feet and legs continued to look healthy, although we had our usual generous quota of wet days. I made sure that the floor of her stable was clean and dry, so that her feet would have the chance to dry out at night. Also I kept the mud in the yard fairly well scraped up so that she could pick her way along with clean feet.

Ñusta was adept at scratching almost any part of her anatomy with her hind claws. She would stand poised on three legs, looking thoughtfully into the distance and pouting her upper lip, while a long hind leg reached the target with extreme accuracy. It looked a very difficult performance, like scratching with a barge-pole, but I suppose it wasn't difficult for her. She scratched the parts of her body which couldn't be reached by a toe with her long lower teeth. The swan neck would curve round over her back and she would give the ticklish area a good raking with them, mouth slightly open. She uttered a sort of gobbling sound while this was going on, and the soft, last inch of her ears would vibrate rapidly.

This scratching made me wonder if she had any parasites. Lice were a problem with the cattle and ponies and we usually had to powder them all twice in a winter. However careful we were, the lice never seemed to die out entirely and I suspected that our animals became reinfected from the old oak partitions and dry-stone walls of our buildings. If they were, Ñusta would be at risk. I examined her skin at frequent intervals but could never find a sign of unwelcome visitors. Much later I learnt that most llama parasites were specific to llamas so that she was unlikely to catch anything on our farm.

I was also on the look-out for ticks. These parasites flour-ished on our land and would infest the sheep, cattle, ponies and dogs in the summer months. The cat would get a few and occasionally even a human would find one attached to the skin. The ticks were carriers of a dangerous fever, called red-water in our district, which affected adult cattle. Calves reared in a red-water district would become immune to the disease but an adult animal, moved from a free area to an infected one, would rapidly become ill. We never had any trouble with this disease—all our cattle were home-bred—but I wondered if llamas might be affected by it. Extensive inquiries about this failed to produce any facts. We should just have to keep our fingers crossed.

We found, however, that llamas did not make a very acceptable diet for sheep ticks. During that summer, Ñusta had just one tick, which attached itself to her cheek, a little below the eye. There were no ticks on the insides of her thighs or in the bare part inside her forelegs, usually favourite feeding-grounds for ticks, just this one on her cheek. I tried to remove it several times but it was rather small and she would not stay still. When at last I succeeded in taking it off, I found that it was a dried-up and poorly-looking tick, not the usual juicy and flourishing parasite. It looked like a tick that had been having the wrong sort of diet.

We noticed that flies behaved differently with Ñusta. They would whirl above her head in a little cloud, but they never swarmed round her eyes or settled on her body as they did with the cattle and ponies. I never saw a cleg, warble fly, New Forest fly or any of the other ferocious biting insects bother her. Perhaps a llama seemed as strange an animal to them as it did to us.

# 12 Llama Drama

Even at the time we chose it, I had never wholeheartedly liked the name of Ñusta. I quite liked it and it was difficult to find anything better. I was also glad that we were using an Inca word. In theory it fitted the llama perfectly but in practice it didn't seem quite right to me. As time passed Ñusta acquired, like many of our other animals, various ridiculous nicknames. These stuck. She learnt to answer to them and they provided us with a good deal of feeble amusement. The main nickname was the Um, because of the little *mmm* sound that she made. This sound also inspired the name Toots, and then Tootie-Frootie. These later grew to Tootles, Frootifer and Frootington. Such names were absurd but they amused us and seemed to fit the animal in an odd sort of way. There was a small corner of my mind that thought it was hilarious to own a llama called Frootington, particularly such a charming, cuddly and delightful one as ours.

The Um would put her face to mine and sometimes we would have quite long umming conversations together. Sometimes her toot was a low and musical note, sometimes it was no more than the merest whisper, just the faintest breath of a toot which was scarcely audible. If she wanted to come into the cottage, she could toot quite loudly. She would go umming backwards and forwards between door and window till someone took pity on her and let her in. At times the note could be quite long-drawn-out and rising in desperation. It was a hard heart that could ignore it.

She was becoming more and more part of the family. She was really getting what Paul called cUMpanionable. If there was anything happening, the Um liked to be in on it. If we were gathering the sheep, she would rush to join in, not mixed up as part of the flock but definitely on the side of the bosses. If I read a story to the children, she would come to sit down in the family circle and listen to it. If visitors in cars arrived, she would hurry along to interview them. The huge eyes and radar ears missed nothing. She liked to be in the very centre of events.

Hay-making was a great pleasure to her that summer. She didn't want to miss a minute of it and was there all the time to oversee operations. She was interested in the mower, but seemed to realize it was dangerous and kept a respectful distance. When we were turning the hay with hand-rakes, she would sit down just ahead of us and be part of the scene. We would have to rake round her and pass on, leaving little llama-shaped patches of unturned hay behind us. She was fascinated by the windrows and hay-cocks which miraculously appeared and transformed the landscape for her. She would hurry round to inspect each one, then, when thoroughly orientated, she would lie down by the heap of her choice and begin to eat it. She seemed to feel that we were making the hay specially for her and would always eat it in preference to the growing grass.

We noticed that she lay down much more frequently than the other classes of farm livestock. Perhaps the slender legs were not designed for a great deal of standing. '. . . they are but feeble animals,' Oliver Goldsmith had said. '. . . they are obliged to repose . . .' She did repose a good deal, but while reposing she also ate. With such a long neck, she could graze comfortably in quite a wide half-circle as she sat. Indeed about half her grazing time seemed to be spent sitting down. If we paused for a rest, she would come over and join us. It was delightful to have an animal that was so interested in our activities and so enthusiastic about them.

The weather was hot that summer, just right for hay-making. When we first came to Carneddi, hay was made on all the small hill farms in the locality, but now we were the only farmers left who made it. I had always believed in self-sufficiency as far as possible and this belief had been reinforced by our Welsh farming neighbours in those early days. They made enough hay to feed their livestock through the winter and, if they were still making it late in September after a wet summer, they did not consider it unusual. They grew a few oats and a few potatoes and what they did not produce themselves they did without. This outlook was the tail-end of an old tradition of self-sufficient family survival in a poor district, and it was a hard one. During the 1950s and 1960s, as the older people died or moved away and the economy altered, this system changed. The farming was done in larger units and supported fewer people. Welsh farmers now bought their hay in Shropshire, or as far away as Lincolnshire or Sussex, and their small, awkward hay-fields went back to rough grazing and bracken.

I was, however, so indoctrinated by the old ways that we went on making hay in spite of the changes. In the summer of 1946, I cut all our grass with a horse mower. For the next few summers, my father and I cut it with the same mower towed behind an old US Army jeep. Later still we had the mowing done by contract and finally Paul bought our own small tractor but, however we did it, the work was always hard. No baler could get on to our small sloping fields. We turned most of the hay by hand and carted it loose to the hay-barns. We probably laboured very hard to get hay worth £12 a ton but, when the price shot up to around £70 a ton in 1974, we felt that our efforts were better rewarded. Whatever the price of hay, we felt much satisfaction in gathering in our own and having it safely stored on the farm to tide us over the winter months.

During that hot summer, we laboured away enthusi-astically with the sweat pouring from us. As old John Williams used to say to me: 'Hay is always better with plenty of sweat on it.' Our hay was good that year. In the middle of the afternoon Ann and John, with their friends, Amanda and Penny, would ride over to the pool at Tanrhiw to swim, leaving the overheated adults to toil on. It was a fine sight to see the cavalcade streaming past at a gallop, with a foal running loose and two or three dogs racing with them.

The pool was in the little Nantmor river, about a mile up the valley. Here, in times past, we had washed the Carneddi flock a few days before shearing. On mountain farms it had been traditional to wash the sheep. Washed wool fetched a better price than greasy, though it lost a little weight. Even a few coppers were important in those hard times, and no one minded working to earn them.

The washing pool belonged to the little farm of Tanrhiw. A few days before we were ready, one of us would walk down to ask Mrs Jones, Tanrhiw, if we might use the pool. It had been used for washing the Carneddi sheep for as long as anyone could remember and the answer was always 'yes'. William Owen, our neighbour at Corlwyni, also washed his sheep there. Now he would go down to the river and mend the dam where winter floods had swept the stones away. There were good stone pens on the left bank and he would also check that these were in order.

When washing day arrived, we would drive the whole flock down to the valley with the help of two or three neighbours, cross the river at the ford and go upstream for a few hundred yards to the Tanrhiw pens. Here the sheep would be passed from hand to hand and thrown into the pool. I stood on a rock which jutted out into the river and ducked the sheep, using a long pole with a T-shaped end, to make sure they had a good washing. Then, as the animals swam strongly through the deep water, I could guide them to the little stony beach where it was easy for them to scramble out on to the bank. Finally, we would drive the now damp and subdued flock home again.

As the years passed the stone pens gradually fell into disrepair. Mrs Jones no longer farmed Tanrhiw. William Owen retired. Then another neighbour built a netting pen on the right bank of the river and allowed us to use it. It was a good pen; the bank was higher on this side and the sheep had a high dive before their swim.

As the years passed, most farmers began to find that the cost of labour to wash the sheep was not offset by the higher price of washed wool. The netting pens on the right bank of the river also fell down and were not repaired. Sheep washing was becoming a thing of the past. We made our own pen in the lower Clogwyn field and a small pool in a tributary of the river. Here we washed for several years. It was not so effective as the big, clear pool at Tanrhiw, but it served us until we too gave up the washing for lack of labour.

But the pool remained a beautiful place for swimming. It was set in a little ravine, fed by a small cataract and overhung by hazel, birch and oak. It was just deep enough for diving if you picked the right place, and wide enough and long enough for a few yards of swimming. Inevitably, some of the tourists found it and, over the years, the good old stone pens disappeared, the rocks cast into the river or used for building dams. On a hot summer afternoon, bathing tourists might be surprised by the Cowboys and Indians from Carneddi who came cantering up to swim in the pool, ponies and dogs included.

One warm day, when the hay-making was almost finished, Paul and I, Becky, the pupil and a friend decided to go to the pool with the children. We felt we had earned it.

'Let's take the Um with us for the walk,' Paul suggested.

So far we had not taken her off the farm, but she followed us everywhere that she could go. She might enjoy some new country. We stowed towels, bathing costumes and some biscuits into a rucksack and set off. The riders trotted on ahead. I had Ñusta on her collar and lead to get her started. Sometimes she loitered behind and I did not want her to miss the way on the parts that were new to her. She followed me meekly.

When we had passed through the wicket gate to Corlwyni, the ground was more open and I let her off the lead. I felt fairly certain that she would follow us. She did. Not only that, she wanted to keep right in the midst of the party. Her normal stately walking pace was slower than ours so that she kept having to put in a few quick galumphing strides to keep up. She seemed very aware that she was now treading new ground and was all agog with interest at everything she saw. As she hustled along at my side, her neck was more than usually upright and the feather duster was carried at a saucy angle. Every now and again she would give a little toot. Her huge eyes did not miss a single detail of the scene.

Her long ears were turned forward to catch every sound. You could almost feel that she was saying: 'Isn't this fun? Isn't this interesting? If I keep close to them I shall be safe. If I go where they go I shall be all right. Let's hurry along and see what's over there!'

It was great fun to take such a well-behaved and enthusiastic animal for a walk. Her obvious enjoyment of the outing made everyone else enjoy themselves more. It was like taking a small child out on some special treat.

When we reached the Gelli, the flat field that bordered the river at Clogwyn, I clipped on Ñusta's lead again. Now we had to cross the river, either going through it or over the foot-bridge, which was two planks wide and had only a couple of shaky wires by way of a handrail. I led the llama to the bridge and she followed willingly till her front feet were almost on the planks. Then she baulked.

'Come on, Tootie,' I encouraged her. 'Come on. It's quite safe.'

I pulled on the lead and she pulled back.

I had read somewhere that when a llama lay down through weariness or overloading, the Indians in charge would 'kneel at its side to encourage it for further exertion by a profusion of flattering epithets and gentle words'. I had already found that Ñusta responded to my voice. Now I pulled on the lead, coaxing her with some flattering epithets.

Quite suddenly the lead went slack and the llama crossed the bridge in a couple of bounds, nearly knocking me into the water. She dashed over with flattened ears, uttering a frightened little high-pitched toot as she went. We both arrived safely at the other side.

'Oh, what a brave creature!' I said.

Once on safe ground again, Ñusta was quite happy, but I found that it was rather touching that she had been prepared to cross so perilous a bridge with me. It was a little humbling to feel that the llama trusted me, wanted to please and to go where I went.

When we reached the Tanrhiw pool, we found that the children had tied the ponies to trees and were already swimming. The adults made ready to follow them. The cool greenish depths looked very inviting after the hot hay fields. Paul and Becky dived in. I stayed on the bank to keep the llama company. She was still a little frightened of the ponies and did not like to get cornered by one of them. Now she seemed very intrigued by the swimmers and went to stand on the big stone where, in times past, I had stood with my long pole to duck the sheep at washing time. I don't think Ñusta had seen any large quantity of water before, except perhaps the sea-lion pool at the zoo. Carneddi and Tŷ Mawr are up on a spur and there are no large streams on our land. Now Ñusta balanced on her little feet on the big rock, bowing her head up and down at the pool and tooting with excitement.

'Be careful you don't fall in, Tootie,' I warned. 'I don't think you are an aquatic animal.'

Then Paul came out of the water.

'Will you look after the Um while I swim?' I asked him.

In spite of the heat of the day, the water in the river was bitterly cold. I never knew how the others could dive straight in; I felt as if I should die of shock if I did so. I began to lower myself in cautiously, inch by inch, enjoying the coldness of the water but hardly able to bear it.

'Come on in, Mum,' said Ann. 'Don't be a coward.'

'*I'm* right in,' said John.

Just then there was a huge splash. While I was occupied in trying to submerge myself, the llama had hurled herself from the rock into the centre of the pool.

'Get out of the way, John,' I shouted.

I knew the danger of the flailing hooves of a swimming animal. The llama was making straight for him, the nearest familiar object in an alien element. She was obviously upset. Whether she had expected the water to bear her weight or whether the coldness shocked her, we could not guess. She didn't appear to be a very efficient swimmer but she came forward in a mass of spray, with her head and neck reared above the water like a miniature Loch Ness monster. John went under, giving the swimming llama no support. She continued to the bank and struggled out with water streaming from her. John bobbed up again and also reached the bank but he was crying. I was out now too. The whole incident had happened in a matter of seconds.

I wrapped a towel round John and comforted him as best I could. There was a long bruised weal down his thigh where one of Ñusta's claws had scraped it. He was weeping with pain and fright and coughing up water. The llama just stood there with a cascade streaming from her sopping fur. With all her fluff plastered down, she looked about half her normal size.

'Well, we know now that llamas can swim,' said Paul.

Bit by bit, John cheered up. He was a brave little boy, and now he began to see how exciting it was to have been swimming with a llama. I was able to turn my attention to Ñusta. She was standing there, looking miserable with rows of drops plopping from her belly and trickling down her legs. Unlike a dog, she seldom shook herself. She did not shake now but stood looking a little surprised and very wretched. Now that she was wet we could see what a skinny greyhound-shaped figure she had. Under all that wool, there was really not much llama.

'I told you that llamas are not aquatic,' I said as I wrung out handfuls of wet hair along her sides. It felt like soggy cotton wool.

After that, we set out on the homeward trek. Ñusta cheered up when we were on the move again and began to pull at rushes and leaves as she passed.

When we came to the river crossing again, the Um seemed very unwilling to attempt another passage of the bridge so I led her through the shallows. She followed me after a little hesitation, with the cold water above her knobbly ankles and her strange feet balancing on the underwater stones.

On reaching home, I found an old dog towel and did my best to dry her but the towel quickly became soaked without making much impression on her wet fleece. That night, when I went to Carneddi to sit with my mother, I invited Ñusta into the warm kitchen. She sat herself down on the red tiles of the kitchen floor, pulled the newspapers from the shelf and began to eat. She didn't seem any the worse for the swim. When it was time to go and the Um stood up, there was a wet, llama-shaped patch on the tiles. It was another two days before her fleece was restored to its normal fluffy dryness, but there were no ill effects from that little llama drama.

# 13 First Birthday

Ñusta was one year old on 11 August 1975. Of course she had a birthday party. We had now had her for nearly seven months, and she had grown from a frail, bony baby into a sort of teenage-looking llama with a well-developed character, a bouncy enthusiasm for life and an established niche in the family. So far the pitfalls of amateur llama-husbandry had been by-passed without our noticing them. We had good cause for celebration.

I made a birthday cake for the occasion and decorated it with Maltesers, one candle and a couple of plastic llamas from the toy-box. I knew the Um wouldn't like it—it wasn't her sort of food, except for the Maltesers—but everyone else would enjoy it. Beenie bought some Cadbury's chocolate fingers for her, Ann bought a brush, John a new lead and Mother picked a special bunch of sweet peas. As always at that time of year, we had a great profusion of blue hydrangeas so I gathered some and twined them into a garland for her neck. Whether she would hate the garland or eat it, I did not know.

In the afternoon Amanda and Penny arrived as birthday guests, bringing with them presents of chocolate and beautifully drawn birthday cards decorated with lifelike pictures of Ums. Our friend, Marjory Barlow, was staying with us at the time. Though she had seen and done much and often moved in distinguished company, she said she had never before in her life attended a llama's birthday party.

I produced the garland just before tea-time. First I let Ñusta touch it with her nose but she wasn't very interested. As far as eating flowers went, she preferred roses and sweet peas. She didn't feel up to tackling a garland of hydrangeas so I hung it round her neck and she didn't object. It suited her beautifully. At any time she was an elegant animal; with the garland she looked like some fairy-tale creature, as strange and mythical as a unicorn. The blue flowers brought out the colour of her eyes and their crisp texture contrasted with the silkiness of her fibre. She wore her wreath with grace and dignity.

Indoors, she trundled round the Carneddi kitchen, scrunching up chocolate biscuits and Maltesers. She consumed her bunch of sweet peas with great enjoyment but she declined to enter the dining-room; it was not her usual territory. We had the party tea without her. Indeed I had prepared it for humans rather than llamas but, in her own sensitive and polite way, she seemed to know that she was the centre of attention and admiration. I think she had a nice day.

Now we had a yearling llama. Besides making a good deal of growth, she had put on weight. Her spine was no longer a sharp ridge but had a good layer of meat and muscle on each side of it. Her neck, which had not felt much more substantial than a bit of fluff-covered fuse-wire when she came to us, was now solid and muscular. I noticed how hard her body felt. Touch plays quite a big part in one's assessment of an animal's condition. After years of feeling all sorts of animals, the llama's body felt extraordinarily different to me, very dense somehow and hard. If I'd been blindfolded, and quite apart from the texture of her fleece, I don't think I could ever have mistaken the feel of her for a horse or a cow. Her skin felt surprisingly tough, too, and I could well believe it when I read that a llama's hide made excellent leather. It was difficult to guess her present weight. Gone were the days when Paul could stand on the bathroom scales with the llama in his arms.

It was fine to see our llama in such good condition now but I began to wonder if she were not putting on too much weight. Her legs were still slender and spindly, and I noticed that she used her hind legs in a decidedly peculiar way. As she walked, her hocks rotated outwards a little with a strange, loose-jointed look, giving a sexy waggle to her progress. It was a gait unlike any I had seen before. She didn't look fat but I wondered if she could be outgrowing her strength. However, she seemed well enough and we were curious rather than worried.

Autumn was coming and it was time to lift the potatoes and find a wintering farm for the ewe lambs. With the beginning of the school term, the influx of visitors to the district had lessened noticeably. This was something of a relief. In summer our weekly Friday shopping expedition to Portmadoc was quite an ordeal. There was nowhere to park the Land Rover in the High Street and we had to make many journeys with heavy shopping bags to some distant back street, the only possible parking place. The pavements were so crowded that it was difficult to walk along them. The shops were so full that one had to wait a long time to be served. The whole business took far too long and, when it was over, we were thankful to escape back into the hills.

The tourists not only filled Portmadoc but overflowed on to the mountains in all directions. A public footpath ran past Carneddi house and was often used by walkers in twos and threes or in large parties. We found it extremely entertaining to observe the tourists' encounters with the Um. A good many of them did not know what manner of beast she was. After all, a llama was not the kind of creature one expected to meet on a Welsh hill farm.

One day Mother was working in the vegetable garden when a couple of walkers passed by. They did not see her because the garden was on a higher level than the path and was screened by a stone wall. The llama was on top of a small hill some distance away. Of course she had noticed the walkers and was standing bolt upright observing them intently. The walkers noticed the llama. Mother heard the man say something to the girl but she didn't catch what he said. The girl's reply was quite clear.

'Don't be silly,' she said. 'Of course it isn't.'

Aha, thought Mother, but it *is*.

Ñusta was fascinated by the passers-by. If she were near at hand, she would hurry up to inspect them. This inspection was a sort of llama-style interview. She liked to look closely into the face of each person—and her head was now nearly on the level of most humans'—and to touch their hair lightly with her nose. Beards were also of great interest. She didn't exactly sniff at people but would gaze at them intently, inhaling deeply for a few long seconds. Afterwards and less happily, she might take a quick look under the clothing of any skirt-wearing lady who happened to be present. People found this embarrassing but it was hilarious to watch from a distance.

If the interviewee was relaxed, interested in llamas, prepared to stand still and not to take the initiative, Ñusta radiated charm and friendliness. Her ears would be forward and her eyes bright with curiosity and enthusiasm. You could almost hear her murmuring: '*Enchantée*,' to her unexpected guests. More often, those interviewed relapsed into shrieks and giggles at the sight of the llama, flinched away with nervous laughs or, worse still, attempted to paw at her. Her beautiful wool was so inviting to the touch that one could hardly blame them, but there are few animals that like to be touched by strangers and fewer still that actively enjoy it. Usually it is the stroker rather than the stroked that gets the benefit of this activity. Ñusta thoroughly disliked the familiar pats which tourists were inclined to bestow on her. Her feelings were also hurt if people were afraid of her friendly advances. She would withdraw to a little distance from the offenders and stand broadside on with her ears hard back, looking at them from one goggling eye. She grew very tall in this attitude, with fluffy tail up and long neck vertical, almost standing on tip-toes. Her cheeks would blow in and out after the manner of a trombone player—'oompahing' we called it. This was not a prelude to spitting. It indicated that she had been offended, that protocol was violated and that the visitors could push off—the sooner the better as far as she was concerned. She never spat at visitors unless they persisted in stroking her long after she had given them warnings that their advances were unwelcome. Occasionally she would give a loud warning snort or sneeze. This was not spitting. It was the same sort of alarm signal that a wild sheep will give when surprised, a sort of whistling sneeze to alert other members of the flock. She would accept more familiarities from children than she would from adults, but she required a certain standard of consideration and courtesy from everyone.

Gradually news of the llama's presence spread around. People who had seen her once came to have another look, and recommended their friends to come. They came, not to see me now, but to see Ñusta. I thought of instituting a Llama Welfare collecting-box to pay for the disturbance but the embarrassment of presenting it quenched that idea, and anyway she was there to be enjoyed.

We arranged wintering for our ewe lambs at a farm in the Lleyn Peninsula. We were surprised when the farmer there said: 'I hear you've got a llama.'

'Yes, we have,' said Paul. 'How did you hear that?'

'The A.I. man told me,' the farmer replied.

He had a fair-sized dairy herd and the man from the Artificial Insemination Centre often came to his cows. Less frequently the A.I. man came to us. We would leave the cow in the cowshed at Carneddi and put everything ready, soap, towel and bucket of water, so that he could carry on with the job if we happened to be down at Tŷ Mawr. Apparently there had been no one about at Carneddi when the A.I. man last called. He opened the boot of his car, put his head inside and began to collect his equipment. Just then he felt a light touch on his shoulder and became aware of someone by his side. He straightened up and turned, expecting to find Paul or me beside him.

'But I was looking into the face of a b—— llama!' the A.I. man had said.

Nothing had prepared him for an eyeball-to-eyeball confrontation with a llama at that moment. He did not know we had one and Ñusta had made her usual silent approach on little leather-soled feet. He got the shock of his life—but I expect it made a good story to tell for a long time afterwards.

I should think Ñusta was the most photographed llama in North Wales, perhaps even in the world. Everyone who had a camera—and most of the tourists had—wanted to take her picture. A snap of the wife and kids stroking a llama on the wild Welsh hills was not to be missed for the family album. She was a superb photographic subject too, with her beautiful colouring and great elegance, in that photogenic setting of mountains and woodland. I am afraid that many of the pictures must have shown her in a 'we are not amused' attitude but, of course, it was not to be expected that holiday-makers could know much about llama psychology. Anyway, I was glad that her presence gave pleasure and interest to the unwary hill-walkers who came our way. One of the purposes of having a llama had been to add a little spice to life and bring a touch of the extraordinary to a generally prosaic world. The more people who enjoyed her, the better.

Just like everyone else, Ñusta was a little self-conscious about having her photograph taken, but she got used to it. Flash photography did not seem to worry her particularly. She would sit in unruffled calm on the hearth rug while Brian took picture after picture of the pleasant scene. She didn't mind the flash any more than she minded John's firing his cap-gun close to her ear.

J. was another dedicated 'llamatographer'. Like all good photographers of animals and children, he managed to fade into the background. Often, on his weekends at Tŷ Mawr, he was somewhere about with his camera, but you didn't notice him. Animals and people forgot to pose, while he captured a series of fascinating pictures of their doings. He took photographs of Ñusta poised on hill tops with magnificent mountain peaks behind—'llamarama' we called these—also close-up shots of her strange feet, fur and face.

It was a good thing that photographs were being taken; I might need them later.

Paul said: 'You'd better get going on writing a book about her.'

'Yes,' I said, 'I'd like to but there doesn't seem much spare time. I don't do half that I want to do now.'

'Well, I think you should make a start.'

So I began to write the book in my head, but that is as far as it got just then.

Sometimes it amused us to talk about the idea as we sat round the fire in the evenings, with the llama, much larger now, hogging the best part of the hearth.

'I think you ought to start,' Paul kept saying.

'Yes, I do,' said Beenie. 'I think people would love to read about the Um.'

'Of course,' said Paul, 'you won't really have arrived as a writer until you are in a position to write the book of the film of the book.'

'That'll be the day.'

'Then there can be an Um Musical, followed by *Llama Farmer on Ice.*'

'What about Um T-shirts and cuddly Um toys?'

'Those too.'

The trouble was that we weren't commercially minded enough to be able to exploit what lay on our doorstep—or rather hearth rug—but it amused us to think about it.

# 14 Bracken and Brandy

In the third week in October we sent the ewe lambs away to wintering. Autumn was really here now. The woods below Carneddi were decked in splendid colours, with here and there a leafless branch amongst the brown and gold. Sometimes the nights were frosty and the fields white by morning. John was in the habit of getting up early to play with his dog before the rest of the family were out of bed. One morning, when there was a very white frost, he came bounding up the stairs.

'It's snewn!' he cried excitedly.

Since his birth, winters had been mild and he had never seen anything more than a light sprinkling of snow. Ann had not seen much more, and now she jumped out of bed and rushed downstairs to look. Only a view of sky and treetops was visible from the skylights of Tŷ Mawr. She returned, disappointed.

'It's not snewn,' she said in squashing tones. 'That's only frost.'

It would soon be winter, but we faced it with a good stock of hay in the barns. We had quite a quantity of good-quality homegrown hay and I had been able to buy a load of moderately priced straw through the kind offices of a fan who was also a land agent. We didn't mind the winter if there was plenty of food for the animals. We had a store of bracken too for bedding for the calves and foals which wintered indoors. Bracken is normally the bedding on hill farms. There is any amount of it for the gathering, and straw is too precious to be put on the floor. It is fed to the animals instead.

Though the rusty brown of the dead bracken looks beautiful where it covers the slopes of the mountains and though it is useful and cheap for bedding, it is really a dangerous weed. It is said to be the most successful weed in the world and covers millions of acres of the earth's land surface, from just inside the Arctic Circle to the mid-Tropics. It contains a substance which inhibits the growth of other plant seedlings and so spoils the pasture where it grows. It can also be poisonous.

There were acres of bracken on our land. We worked hard to keep it in check round the hay fields but it flourished rampantly on other parts of the farm, sometimes growing to six feet in height. The method of control was to cut it, or slash it down with sticks in June when the strong stems had pushed up through the ground but before the fronds had unfurled. This cutting weakened the underground rhizome and had to be done again in September. If this twice-yearly cutting were done for three consecutive years, the bracken was so much weakened that it almost disappeared—but not quite. It seemed always to be there, ready to pop up and regenerate if the cutting were again neglected. Most of the work had to be done by hand because of the rocky and steep nature of the ground. There had been much less bracken on the hill farms in the old days. The farmers and their families had worked diligently to keep the land clear but now, with fewer farmers and high-priced labour, the green tide had crept back again.

Then aerial spraying seemed to be the answer, but it was expensive. There was a Government grant to assist farmers with this, but the farmers still had to find hundreds of pounds out of their own pockets. Before scraping round to collect some money, Paul and I waited cautiously to see the actual, practical results of the spray. The correct time to treat the bracken, we heard, is in August when the fronds are senescent. We watched with great interest when our neighbour's land was sprayed one August. The children were thrilled to see a low-flying helicopter so near and rushed to watch. I was a little worried to hear that they, also, had been sprayed but fortunately the spray did not prove to be fatal to children. It wasn't fatal to the bracken either. The next year, strong green shoots appeared on our neighbour's land as they always had done in the past. There were patches, here and there, that seemed to have been killed or weakened but, by midsummer, the green sea looked much the same as it always had done. By the following spring it was impossible to tell that the land had been sprayed at all. There was talk that the spraying had been inefficiently done but, by then, the aerial spray firm had gone bankrupt. Paul and I were glad we had waited.

Cutting by hand still seemed the most effective way of keeping our bracken under control. We hadn't the time to cut more than a small proportion of it but the area mown was greatly increased by the efforts of the boys from the Oxford and Worcester College Boys' Club and pupils from Peers School, Oxford. In the early 1960s, Paul's old army friend, Les, who was a youth leader, had brought the first party of boys to camp on the farm. Les believed not only in service to youth but also service from youth, and the lads alternated a day's mountaineering with a day's farm work. They dug ditches and cut wood and bracken, skills which they soon learnt and which they seemed to enjoy. It was of the greatest benefit to us to have an army of willing workers appearing three or four times a year to attack these jobs with energy and enthusiasm. The leaders, Les, Mick and Julian, assured us that the work and the wild, beautiful surroundings benefited the youngsters too. I hope it did. Their mothers certainly had some very muddy clothes to wash on their return.

Besides ruining the pasture, bracken is poisonous to livestock. Generally it doesn't seem to be very palatable but animals do eat it. A dry spring, when the bracken grows and the grass doesn't, can be a time of danger. Young animals seem more liable to be affected than older ones. In thirty years we had lost perhaps five calves with bracken poisoning—five too many, I felt. When the first symptoms appear, bleeding from the nostrils and blood in the dung, it is too late to do anything and the calf dies. These deaths I regarded with sorrow and as a failure in management. They had happened in years when we had run out of hay too early in the spring and had turned bucket-fed calves out to pasture. I found from experience that it was quite safe to turn calves out, so long as they had a good supply of hay until the bracken had reached a mature stage in its growth. Suckled calves did not seem to be affected by poisoning.

Now, with a precious llama on the farm, I felt it would be advisable to try to find out more about the mysteries of bracken poisoning. Sad as it was to lose a calf, it would be very much worse to lose our one and only, charming and valuable llama. If we knew more about the poison we should be better able to protect her from it.

Paul and I had heard that research into bracken poisoning was being carried out at the University College of North Wales at Bangor. We made inquiries and finally made contact with Dr Antice Evans of the Department of Biochemistry and Soil Science. She very kindly devoted the best part of an afternoon to demonstrating as much of her work as our unscientific and untutored brains could grasp.

In a way, that afternoon at the University was a poignant occasion for me and one I shall never forget. We had just taken six-year-old John to the hospital for an ear-drainage and adenoid operation, in the hope it would relieve his partial deafness. We were to see him again for half an hour that evening. It was the first time that he had been in hospital, and it was the first time that we had been the parents of a child in hospital. He was a brave little boy but, like many children with a hearing loss, he found the world more hostile than a child with full hearing found it. His mother was vitally important to him but hospital regulations did not allow me to see him the following day, the day of the operation. He would be alone among strangers in a strange place, where he would be subjected to unpleasant medical procedures. We had explained everything to him as carefully as we could and his acceptance and stoicism were almost more than I could bear.

As we went round the laboratories, my heart was with John in the hospital only a few hundred yards away. My mind was considering his problems and my spirit was trying to strengthen his. The sight of dozens of tiny Japanese quail, each behind bars in a separate cage, helpless victims of research, made me think of the helpless child we had just left in hospital. Ridiculous nonsense, I knew, to compare the two; ridiculous to be upset by either. John was in hospital to be made better; the quail were in their cages to alleviate death and suffering in the future. I made an effort to concentrate on the bracken.

Dr Evans was most interesting and informative. We learnt that bracken belonged to a family of plants which had existed on Earth for 150 million years. She showed us a sample of bracken which had been dug up on a Roman site near Hadrian's Wall and which had lain there for perhaps 1,800 years. It looked just like any bit of bracken that I might fork out of the bottom of the calf pen at the end of a winter. It was pressed flat but the shape of the fronds was clearly preserved. It was interesting that the Romans had used bracken for bedding too, but the specially interesting thing about this particular fragment was that it had been found with a number of stable-fly larvae on it. Why had they never hatched out? Dr Evans told us that there are compounds in bracken which have a marked biological activity and she suspected that these had inhibited the development of the insects. When I came to think of it, I had never seen caterpillars, aphides or even slugs and snails feeding on the plant.

I had not realized till then that bracken could also cause cancer. Shikimic acid had been isolated from the plant. This had been demonstrated to be a carcinogen for mice but there were other cancer-producing factors too. The symptoms of acute bracken poisoning were found to be remarkably similar to radiation sickness. It was good news to hear that Dr Evans had had some success with developing a cure. She told us, however, that at present this was too expensive to be commercially useful.

Much of what she said confirmed our own experience of bracken poisoning. Cattle were particularly susceptible to it and young animals more so than older ones. The early June bracken fronds were the most dangerous and toxicity declined as the plant matured. In countries where bracken shoots were eaten as a vegetable, the incidence of stomach cancer was high. Dr Evans thought a stress factor was usually involved as well and that, if an animal was happy and thriving, it was less likely to suffer from poisoning. She did not know how susceptible the camel family, including llamas, were to the poisoning. It was something that time might show us.

We left the University, very grateful for the help and courtesy we had received and feeling much the wiser. We paid a last quick visit to John, said good night and left him to his fate. He was very brave. That night I dreamed of little Japanese quail and children in laboratories.

In three days John was home again, seeming none the worse for his experience but, alas, none the better. His hearing was not improved. It was the beginning of a long effort to find the best doctor and the best treatment for him, also the best means of helping him to overcome the disability. He was an intelligent child and in time we made progress.

After our instructive visit to the University, and during the following spring and summer, we had kept a very keen eye on the llama's bracken consumption. At first she ignored the springing green shoots, then, as the fronds began to uncurl, she would take a quick snatch at them. They were now at their most toxic stage of growth. At that time she ranged over quite a limited area, following me up and down between Carneddi and Tŷ Mawr. There was not much bracken near either house but it grew at the base of the walls on each side of the path. I hurried up and down many times a day and I took to snapping off the new shoots as they appeared. I did it on the move and it hindered my progress by only a few seconds. Soon this began to make an appreciable difference to the amount of bracken that was temptingly in the llama's usual path.

I found that I could prevent Ñusta from eating bracken by shouting: 'No'. She would hesitate, look at me doubtfully and then move on. She seemed to develop quite a guilty conscience about it after a while and if I only looked at her when she was eating some, she would put her ears back and pull peevish faces at me.

I had no objection to her eating nettles. There were quite a few untidy clumps of them growing near the deep-litter henhouse and on these she feasted with evident enjoyment. I was surprised that they did not sting her mouth but she showed no sign of suffering. Nettles are good food but sheep, cattle and ponies will not touch them green, though they like nettle hay. I have seen ponies pawing up the thick, yellow nettle roots in a hard winter and eating them but it seems that only llamas and, perhaps, goats will eat the green plant.

During the early part of the summer, when the bracken was most toxic, we put Ñusta into the stable every night. I felt that if she had twelve hours in the twenty-four when she could eat no bracken, only good hay, the risks would be considerably reduced. When the autumn came and the bracken was red-gold over the hills, we knew we had been successful in our efforts to safeguard her. Next summer she would be more mature and less at risk, and she was certainly contented and thriving.

As the winter advanced, I began to think more about writing the llama book. The years since Ann's birth had been tantalizingly sterile from a literary point of view. It was nearly ten years since my last book had been published, though I still had a trickle of fan letters asking: 'What happened next?' But how could I begin to write without the rest of the family and the farm being neglected? We had Becky that year, one of the irregular succession of pre-college farm students who came to us for practical experience. Even with this extra labour, there seemed to be so many things that nobody but I could do. I knew the old saying about nobody's being indispensable but I didn't think it was true in every context. I was such a slow writer that it would take me years to write another book but I began to ponder ways and means. The most obvious were: be more energetic; be more efficient; neglect everything just a little; cut out some activities—but you couldn't cut out children with all their diverse requirements, and they were my biggest distraction. Still I pondered and then I bought some loose-leaf paper. I made a slow and tentative start but I didn't get on at all well because it was nearly Christmas.

The Christmas preparations were always specially hectic for us because of the turkeys. We had reared turkeys in our brooder house for a good many years now. At one time we used to produce seventy or eighty for the Christmas market but, when the price of feedingstuffs began to rocket, we reduced the numbers to between twelve and twenty-five. It was unnerving to be paying hundreds of pounds for the feed and not to know how much we would get for the birds at the end of it. There always seemed to be an uncomfortable time-lag between a rise in food prices and a rise in the price of turkey meat. We couldn't produce any turkey food ourselves, except kale and skim milk, so we were very much at the mercy of food manufacturers. However, the thought of a tasteless, mass-produced bird of doubtful age and origin did not tempt us when we were used to a delicious, home-produced one. We continued to rear turkeys for ourselves and a few discerning private customers.

The last few days before Christmas were always a frantic rush of killing, plucking, gutting, weighing, labelling and transporting turkeys, besides the more conventional activities. We didn't enjoy it, but it seemed worthwhile when we made a modest profit at a time of year when money was scarce and, at the same time, provided the centre for a sumptuous banquet for ourselves. Of course all this left no time at all for writing but I was resigned. I was getting all too used to that state of affairs.

Miles and J., those excellent honorary uncles, came for the holiday as they had done for several years. Their presence added much to the happiness of the occasion. This year J.

brought a big bottle of cherry brandy with him. We thought that Christmas Eve was a suitable moment to open it. The outside work was all done. Supper was eaten. A big log fire blazed in the grate. The whispers and giggles from upstairs had subsided and I thought that the children must be asleep at last, oblivious now of the empty stockings hanging from their beds.

'Let's have some of that cherry brandy,' said Paul.

Everyone agreed.

He fetched glasses and filled them. It was a warm, Christmassy scene in the room, with the wild and lonely mountains outside and paper garlands and a Christmas tree within. Even the llama looked festive, with a necklet of sparkling green tinsel which Ann had hung round her neck earlier in the evening. Ñusta had come in, as she usually did, to have a quiet sit with the family before she went out to the stable for the night. She seemed to enjoy these evenings with us. After dark, a white face would appear at the window and she would toot to be let in. She would have a small feed of pony nuts in the kitchen and then settle herself quietly in the living-room for the rest of the evening. She was now rather large to occupy the hearth rug—indeed she filled it from end to end and, when she was there, there was not much room for anyone else. She preferred this place in front of the fire and the thicker rug, but generally I barred her way with a chair. Then she would have to be content with the mat behind the sofa. Here she would sit, sometimes chewing her cud contentedly, sometimes just sitting with her tall neck upright and a far-away look in her eyes. Sometimes she would stretch her neck out along the floor and go soundly to sleep.

This Christmas Eve she was enjoying her hour or so indoors. We had been pleased to find that Ums did not eat Christmas trees, but they did eat foil-covered chocolate bobbles so we didn't put any on the tree.

'Cheers!' we said as we sipped the cherry brandy.

'Merry Christmas!'

'Good health and long life to the Um!'

Ñusta raised her head. She had been sitting like a statue for a long time and we had almost forgotten she was there. Now she stood up and crossed the room to me, ears forward inquisitively and eyes bright. She stretched her nose out to investigate my glass.

'That's only cherry brandy,' I told her. 'You wouldn't like that.'

I held the glass for her to examine. She was such a clean and fastidious animal that, if she bit half your biscuit, you didn't mind eating the other half afterwards. Now her double upper lip elongated, like two little trunks, till they touched the rim of my glass. Then, suddenly, slurp, slurp, she was sipping my cherry brandy with gusto.

'Hey, Um,' I cried, 'you can't have that!' and held the glass above my head.

'Oh, give her some,' said Paul. 'It is Christmas!'

Quickly, he poured a small tot into a good-sized bottle top and offered it to the llama. Our seldom-used crystal liqueur glasses seemed rather vulnerable but the bottle top was just right. Slurp, slurp. Cherry brandy was just what Ums liked. Her usually dignified and polite manner disappeared. Her lips reached out. She pushed and craned. Cherry brandy definitely switched her on. I had not seen her so excited before.

'Hey,' I said, 'what does Goldsmith say about a llama's being "moderate in what it drinks, and exceeding even the camel in temperance"? I don't call this temperance!'

'More, more,' said the llama.

Paul gave her a drop more. She finished hers before we finished ours, and we had to drink up with our backs turned to her and take the glasses away to the kitchen. When all traces of the delicious stuff had disappeared, she settled down on the mat once more with a satisfied expression on her face. All signs of her eager frenzy had melted away. She was again dignified and the perfect lady, but we had seen a new side to her character.

'Fancy, an alcoholic llama!'

'She's a tiddly Um!'

But we were glad that she had found that Christmas was convivial.

# 15 Beginning the Second Year

When the New Year came, we had had Ñusta for almost twelve months. After Christmas there was some dry, clear, frosty weather. At night the stars sparkled from a clear sky and the ground froze—Andean nights, we called them— and we let the llama sleep out. She gloried in the cold, cloudless weather and would settle on the frozen ground in her tea-cosy position, looking like a great puff of pale fur in the starlight, a warm animal in the most effective, highly insulated, thermal winter woollies.

On one such cold night, I was enjoying a hot bath before going to bed. It was a luxury which I still appreciated. Until Ann was born, Paul and I had lived the simple life at Tŷ Mawr. For years we had carried all our water with a yoke and buckets from the spring, and had bathed in an old-fashioned hip-bath in front of the fire. It was very pleasant but it wasn't so good as soaking in a full-sized bath. In those early days we had a chemical closet in the shed outside and our lighting was provided by oil lamps. Then, just before Ann was born, electricity had come to the district and Paul had plumbed water into the cottage. He installed a bath upstairs and built an outside WC which would later be an inside one when he had time to make a doorway through the gable wall and build a connecting passage. We enjoyed these mod cons but at the same time we rather regretted our departure from the old ways.

Now I lay in the bath day-dreaming. The children were sleeping peacefully. Paul had gone out into the frosty night to take the dogs for a last run. He seemed to have been gone for rather a long time. Then I heard a faint bumping sound. Perhaps he was trying to fix the cistern of the lavatory. It had given a little trouble recently. Once Paul had found a dead beetle blocking the pipe. Perhaps the same thing had happened again and that was keeping him. I wallowed luxuriously in the bath and the bumping went on. It seemed to be taking him a very long time to mend the lavatory. Then I thought I heard a faint shout. I listened. Yes, there it was again. Something must have gone wrong. He must need help. I couldn't yell 'Coming!' for fear of waking the children but I jumped out of the bath, hastily towelled myself, cast on a dressing-gown and slippers, ran down the stairs and out into the night. Round the corner, I almost bumped into the llama who was standing outside the lavatory door. Paul was in there. I heard him chuckle.

'The wretched animal has locked me in,' he said.

He was soon released and we had a good laugh. He had not been mending the cistern and the bumpings had been made to attract my attention. The llama had gone with him to the lavatory, as she often did, and had waited outside for him. It was her habit to scrape her teeth on the wood-work of the door and in doing so, she must have moved the turn-button which secured the door on the outside. None of his entreaties could get her to open it again: he was well and truly locked in. He was unwilling to force the door because he was the person who would have to mend it again. I was sorry to have been so long in realizing what had happened, but the circumstances were unusual.

'"O dear, what can the matter be?" ' we hummed as we went back into the house.

'You're unique,' I said. 'I don't suppose there's anyone else in the world who's been locked in the loo by a llama.'

'Do you think it's one for the *Guinness Book of Records*?' said Paul.

'Sure to be.'

During the cold weather, we noticed how thick Ñusta's fleece was growing. When she stretched her neck upright, you could notice the rings of pressed fur round it parting very slightly to give a curious corrugated effect, rather like a huge radiator hose. There was a delicate halo of single long hairs over her whole body, neck and head, which gave a kind of gauzy appearance to her silhouette. I noticed that now she had grown a little, silky, white mane along the crest of her neck and that an odd tassel had appeared on the end of her tail. The ginger feather duster, that busy and expressive appendage, had at first consisted of a bush of fur, finishing in a tidy sort of fringe halfway down her behind. Now a single pale-coloured lock had grown from it. The lock, slightly curling, hung down centrally at least three inches below the tidy fringe. It looked most odd. It invited you to snip it off level with the fringe, but we didn't because it was such a strange decoration.

So far Ñusta had shown no sign of moulting. I brushed her a little on most days and a small quantity of fibre came off on the brush, but it didn't amount to much. I saved these brushings and they accumulated slowly until I had quite a big bagful. Of course we were longing to see the children dressed in llama jumpers but that day still seemed a long way off. We had been given a spinning-wheel on indefinite loan, but neither Beenie nor I knew how to use it properly. I had done a little spinning at Oathill Farm when I was a pupil there long ago, but not enough to become proficient. What little I had learnt seemed to have been forgotten and I ended up with a lumpy and overspun thread when I tried to use our wheel. Both Beenie and I were enthusiastic about the idea of spinning but we never seemed to find the time to learn. Spinning is not the sort of skill you can pick up in half an hour. We knew we should have to set about the task seriously, read the right books, make sure that our wheel was properly adjusted and practise on sheep's wool before we attempted to spin the precious llama fibre. Just then, we hadn't the time for all that. I was delighted when Mrs Arnett, a friend in Beddgelert, offered to spin a trial amount of llama fibre for us. She spun beautifully and I had already admired her work. I quickly supplied her with a quantity of the llama's fluff and looked forward to seeing the results.

Some time later, Mrs Arnett returned the finished product. It was beautiful—an ounce or so of very soft, very light 2-ply, of a delicate, light camel colour. It felt much silkier than ordinary wool and I thought it would knit up beautifully. Mrs Arnett said she had found it difficult to spin. There was not so much crimp in the individual fibres as there was in sheep's wool and she found they wouldn't spin until she had oiled her hands slightly. We had supplied her with brushings and she had quite a job to tease the fibre from the rubbish—bits of grass, hayseeds, soil and leaves, for the Um loved to roll, but despite these difficulties, she had made a fine job of the spinning. Paul was so delighted with it that I think, if it had been summer-time, he would have sheared the poor Um there and then to get a further supply of the raw material.

There was a small, rocky hill behind the deep-litter hen-house and here Ñusta liked to sit. If the weather was good, she always preferred some high eminence where there was a commanding view. From this hill, there was a fine view down the valley to the sea and she was strategically placed to hurry down to interview any passing walkers on the footpath below. She could also see the Carneddi front door and would sit facing the house so that she could watch my movements and be ready to accompany me back to Tŷ Mawr. The short turf and small heather plants were much to her taste and she would graze in a half circle as she sat. This hill was called Bryn Eithio which, I was told, meant

Winnowing Hill. I suppose that in times past, when corn was grown on all the small hill farms, this was the place where the corn had been winnowed. The site was certainly exposed and breezy. It gave me great pleasure to see Ñusta sitting there. It was still surprising to look out and see the silhouette of a llama on the opposite hill. Against the sky-line, she made a slender, bottle-shaped figure, surmounted by the two curved ears. I never became indifferent to the sight of that elegant form. I wondered what Carneddog, writer and former owner of Carneddi, would have thought of such a view from his window. Bryn Eithio now acquired the more topical name of Bryn Um.

Whenever we went on a suitable walk now, we took the llama with us. She loved any sort of expedition and would bustle along with enormous enthusiasm. Her normally slow and stately walk would quicken when she was on new ground. Mrs Cole, our neighbour at Corlwyni, had said that the llama's walk reminded her of an ostrich. It was a good simile. The long neck, small head and jauntily held feather duster, which somewhat suggested a small ostrich feather, added to the likeness. There was, however, no hint of an ostrich's ungainliness in Ñusta's movements.

Sometimes we took her for a stroll in the forest which bordered the lower boundaries of Carneddi and Tŷ Mawr. This had been private woodland but was taken over by the Forestry Commission a few years after we moved to Carneddi. We had been sorry to see the deciduous trees felled and the land replanted with conifers but, as the trees grew, the effect was quite pleasant. The conifers were of mixed varieties, including larch. A few beeches and oaks had been spared and a new generation of birch, ash and rowan were springing up. Now the forest was quite dense again and made an exciting place for expeditions for the children, dogs and llama. Ñusta kept close to me in the forest. She shouldered her way bravely through the twigs and branches, keeping a sharp look-out for pumas and snatching at the leaves as she passed. She seemed to have lost the cautious clumsiness of her first months with us, and her prison mentality had long since gone. Now she bounced confidently down rocky slopes and clambered up steep banks. She was not as agile as a goat—and, of course, she was much bigger—but she could go in places that were too steep and rough even for our clever mountain ponies. Paul was particularly pleased about this; he regarded her as a potential pack-carrier for mountaineering expeditions. The ground that did not suit her was marshy, wet land. It was clear that her feet were not designed for bogs and she distrusted them. She would pick her way round the mud with the delicacy of a cat.

I no longer had to lead her. She came with us very willing-ly wherever we went, though I usually had the collar and lead with me if we were likely to go on the road. We always avoided the roads in summer-time. In winter an occasional car might pass and then it was sometimes alarming to see the driver doing a double-take at the strange beast he had just passed.

I felt Ñusta really needed a leather head-collar if we were going to use her for carrying packs or lead her in traffic. Our smallest foal slips were too big for her dainty head and we should have to make one. When we began pony breeding, we had tried our hands at making bridles with great success. The job was quite easy if you had a pattern to follow and we found that we could get a good professional finish on the work. The main ingredients were good leather and time. We had the leather but time was short and so, for the time being, Ñusta followed us unrestrained.

As part of her training, during her first summer, we had laid our coats over her back and she carried them for us in a dignified manner. Then, as she grew bigger and bouncier, she decided we were taking liberties. She pranced to shake the coats off. If they didn't shift, she lay down and rolled. I could see that we should have to do some careful training before she was a reliable pack-carrier. But there was no hurry; she was still young.

It was a strange sight to see the llama playing. Most of the time she was quiet and dignified, but suddenly she might throw decorum to the winds and leap in the air like a gambolling lamb. Sometimes she would stand up on her hind legs and leap. She was about eight feet tall when she did this and it was an impressive sight. She did it seldom and its occurrence was so unpredictable that, try as we would, we could not capture it on film. The nearest Paul got to photographing the giant leaping llama was when I had spread a long cotton mat out on the grass in front of the cottage. I was about to roll it up, ready to take it to the laundry. The llama noticed what I was doing and came hurrying along to see.

'Curiosity killed the cat,' I told her but, as far as she was concerned, mats were for sitting on, so she sat on it. Paul was amused and brought his camera. After sitting for a few minutes, she rolled and Paul photographed that. The long striped mat on the grass evidently pleased and excited her. She stood up and then, all of a sudden, she leaped. Paul pressed the button. Later we found that he had got the picture but, unfortunately, he had not been expecting such fast movement. The speed of the exposure was not fast enough and the figure in the photograph was a huge fuzzy blur, hurtling upwards; only the back legs were in focus.

The llama had another game which could be disconcerting, particularly as she grew bigger and heavier. She would walk up behind me, very silently, rear and throw her weight forward. Her front legs were always well tucked out of the way. She did not hurt, but made me stagger if the assault were unexpected. I became adept at hopping out of the way when I sensed that Ñusta had stolen up behind me and was feeling in that sort of mood. I could detect when there was an I'm-going-to-bounce-on-you gleam in her eye, and it was easy to push her off balance if she did rear up. Fortunately she never tried this trick with strangers and seldom with anyone but me. It probably had a sexual origin, and I was certainly the object of her affections. The bounce was also used aggressively, but we were not to see this demonstrated until later in the spring when Mr Widdle came on the scene. I never felt that the Um was being aggressive when she tried to bounce on me. Her ears were always forward and the action seemed to be a kind of game or love play. I could generally make her do it to order by patting her on the chest, especially at a time when she was feeling skittish. The children thought it was hilarious to see me larking about with the llama, as I sometimes did when we were both in the mood. However, although she was now a big animal, I never found that she was in the least unmanageable. She would always take 'no' for an answer and give in straight away, if I were firm.

With spring on its way, I began to look out for signs of Ñusta's coming on heat. We knew nothing about a llama's breeding cycle except that the gestation period was eleven months. This was the same as a mare's and we supposed that, like a mare, she might come on heat about every three weeks in the spring. I felt she was now mature enough to show signs of oestrus but I saw nothing. There seemed no sign of any change in behaviour at certain times. I watched her closely but the weeks passed and I could detect nothing. I asked our vets if they could tell me anything about it but they did not know.

A cow comes bulling every three weeks unless she is in calf. It is important for cows to be in calf at the right time, otherwise they are just expensive passengers on the farm. Some animals will go bawling all over the place when they are bulling, jumping on the backs of their companions and making so much fuss that only a half-wit could fail to notice. Unfortunately our cows tended to be very discreet about the business and their oestrus was very difficult to detect. We had no bull to take the responsibility and had to rely on our own observations and the excellent artificial insemination service. Sometimes Beenie and I said that we needed to gaze into a crystal ball before we could tell if our cows were bulling. Often we depended on the smallest clues, but we were generally successful. With Ñusta, I felt we should need more than a crystal ball to tell us anything.

Paul and I did not plan to try to breed from the Um that spring but we wanted to collect information. Many people asked if we were going to breed from her. The answer was that we felt it would be a pity to have a beautiful female animal and not to breed. I was longing to see a new-born llama and felt sure it would be exquisite. Whether Ñusta would have other ideas remained to be seen. We felt sure that we could sell any offspring very easily unless we wanted to keep it. We thought that next spring might be the right time to try. We wanted to feel that Ñusta had completed most of her growth before we bred from her. In the meantime we must locate any available males.

Bit by bit, we must assemble the pieces of a jig-saw puzzle of llama knowledge—but there were a great many pieces that we had yet to find.

# 16 Some Solid Facts

Reading was our main relaxation, hobby and, perhaps, vice. Paul, Beenie and I would read our way through most meals with our books balanced against the Golden Syrup or milk jug. Of course we did not read when we had visitors or at Saturday and Sunday dinner when the children were at home, or indeed in the school holidays. We felt that we should be setting such an appalling example and anyway, with five of us reading at the table, somebody's book or magazine would be sure to find its way into the butter or the gravy. We confined our reading at meals to the times when there were just the three of us. There wasn't much point in keeping mealtimes for conversation. Mostly we had been working together all day and had had plenty of opportunity to say all we wanted to say. We read solidly in the evenings too. I quite welcomed the time when I had nothing to read or my book was boring; then I did more useful jobs, like writing letters or mending. We were able to get a good supply of books from the Public Library in Portmadoc when we went shopping on a Friday, or from the travelling library which came to Nantmor every other Monday.

I searched the non-fiction shelves for any books on llamas, in an attempt to improve our knowledge, but there was little that the library could offer. The best I could find was a couple of books on animals in general, each with a short paragraph on llamas. Even here, the information was partly conflicting and very limited. Our best authority so far was still Goldsmith's *Animated Nature*, published two hundred years earlier and written by a man who had, perhaps, never seen a llama.

It was time to try to collect some more solid information. I asked Miss Fargher, one of the vets in our group practice, if she could help me. I felt I needed some facts to go into the book, if I ever got it written, and also some real knowledge might prevent our making some terrible blunder or omission through ignorance. Miss Fargher was a very good vet. She was kind, charming and knowledgeable and the animals loved her. We were very pleased when she said that she would be glad to help us. She said that she found the llama an attractive and interesting animal, though outside her experience so far, and she would gather what information she could in her spare time.

The first books which she found for us, *Wild Animals in Captivity*, the *UFAW Handbook on the Care and Management of Farm Animals*, *Royal Natural History* (Volume Two) and *The Tropical World*, were enormously interesting to us. I enjoyed reading the parts about the other animals too—how to build a hippopotamus pool and so on. We learnt that properly Ñusta should be classified thus:

| Class | Mammalia |
| Order | Artiodactyla |
| Sub Order | Tylopoda |
| Family | Camelidae |
| Genus | *Lama* |
| Species | *glama* |

'I like *Lama glama*,' said Paul. 'She is rather "glamarous".'
We all agreed.

We also learnt that the camel family had originated in western North America from a pig-like animal and had developed quite independently from the true ruminants. The descendants of these animals migrated westwards into Asia and southwards into South America. They gave rise to the two camel species of the Old World, the bactrian and the dromedary, and the four camel-like animals of South America, the vicuna, the guanaco, the alpaca and the llama. The fossils of giant llama bones had been found in eastern South America by Charles Darwin on his voyage in the *Beagle*. Some authorities thought that the llama was descended from the wild guanaco, but others believed that it was the domesticated survivor of a distinct species which had once inhabited the Andes and was now extinct.

The animal had been domesticated since before the time of the great Inca civilization and used for transport. Only the males were used for this purpose and would carry a load of 100 lb. We read that a single Indian would drive several hundred loaded llamas, carrying gold and silver from the mines. With a lead animal, they would traverse narrow mountain paths in single file, encouraged by calls and whistles. The Indians would decorate the llamas' ears with ribbons, hang little bells round their necks and always caress them before placing any burdens on their backs. The animals needed gentle treatment, we read, with regular food and rest, and they might refuse to move with loads of over 100 lb. They would only allow themselves to be loaded if they were one of a group.

The Spanish conquerors of Peru regarded llama meat as being equal to the best mutton and they opened shops in the towns for its sale. The Indians made rugs, ropes and clothing from the fibre, sandals from the hide which was considered very durable, had the milk and meat for food and used the droppings for fuel. It sounded as though the llama was one of the best package deals in history for meeting the requirements of man.

One of our authorities said that the llama had been developed for riding. Another said it was never ridden. All agreed that only the males were used for work.

'That's probably because they had plenty of males and they were stronger,' I said, 'but I don't see why our Um shouldn't carry small packs.'

'*Mmm*,' said the llama from her place on the hearth.

The colour of llamas, we read, varies from black through shades of brown to pure white, with sometimes pied or spotted variations. Where possible they are now being replaced by horses, mules and motors as a means of transport, but they still form the Indians' main source of wealth. At the present time there are nearly a million in Peru, two and a half million in Bolivia, seventy thousand in Chile and fifty thousand in Argentina. Their natural habitat is between 11,000 ft and 17,000 ft on the ranges of the Andes. When frightened or angry, they spit and kick and their bites can be serious because of the long, sharp canine teeth. We read the above facts but they didn't entirely tally with our experience. We knew all about the spitting but we had never seen Ñusta kick. She sometimes stamped irritably if the dogs were milling round her feet but she never aimed a deliberate kick in the manner of a cow or pony. She didn't bite either. She didn't behave like an animal which bites and, so far as I could tell, she had no canine teeth. Perhaps those would come later.

In so far as the books mentioned the subject at all, the feeding and management which we had given to Ñusta corresponded with what was suggested. One would not expect to find a mention of Maltesers, cherry brandy or sugar in those books. It was also said that llamas make good pets and become very attached to their owners—a statement which our own experience fully supported.

One very useful fact which emerged from our reading was that llamas suffer from the clostridial diseases common in sheep. These include the killers like tetanus, braxy and blackleg. With a disease like blackleg, a sheep could be well one day and dead the next. It would be too awful if such a thing were to happen to our precious Um. Paul and I hastened to give her 2 cc of Tasvax 7-in-1 to provide immunity, just as we did for the sheep. Again I noticed the toughness of her skin as I gave the injection. She collapsed on the ground at the indignity of it all but was soon up again and scrunching Maltesers as a reward. A second dose would be required in a month.

We were pleased to read that llamas did not suffer from foot-and-mouth disease, though they were liable to most other diseases of ruminants. There was no mention of red-water fever, the thought of which had worried me, but now, having had Ñusta on our mountainside for a year, we supposed she must be immune to it.

We read that three hundred llamas had been imported to Australia but that, in five years, they had dwindled to twelve, also that an attempt to introduce a herd to the Scottish Highlands had failed. What went wrong, we wondered? It would have been fascinating to know more details of these ventures.

The books gave the llama's gestation period as eleven months, a fact which we already knew, but they had nothing to say about oestrus, or anything detailed about breeding. One book said that births were usually single, another that twins were unknown. In the natural habitat the breeding season was from December to March. That was in the Southern Hemisphere, so would it be different in the Northern? we wondered. We read that young llamas are usually born in the morning so that they can dry out in the sun before the evening drop in temperature, which is considerable in the Andes. Those born later rarely survive. Llamas breed freely in zoos all the year round, we read, but would it be the same on a farm where we had to take the animal to the male at the right time? How would we know when it was the right time? It was clear that we needed much more detailed information about the breeding cycle of llamas.

Miss Fargher now said that she would write to the Library of the Royal College of Veterinary Surgeons and ask for a search to be made for the information we wanted. This proved to be most fruitful. The Librarian sent a list of fifty-four references to the Camelidae of the New World, going back to 1969. She regretted that there were so few available, but fifty-four seemed quite a lot to me. She said that we could have photocopies of any of the articles that interested us.

With Miss Fargher's help, we picked the fourteen that seemed most useful. 'Oestrus and mating behaviour in the Llama (*Lama glama*)' sounded just what we wanted, also 'Llama reproduction—a South American problem'. We decided upon 'The effect of different levels of simulated altitude on $O_2$ transport in llama and sheep' and 'Relative weights of the right ventricle of the heart in alpacas and llamas at high altitude and at sea level' and a few others on the same subject. Besides wanting to know about the breeding of llamas, I was enormously interested in that specialized physiology which made it possible for them to live at such high altitudes. Whether I should be able to understand it remained to be seen. We chose 'The legendary "camels of the Andes" and their production of textile fibres' as being of general interest, but 'Volatile fatty acid concentration and pH of llama and guanaco forestomach digesta', 'Muscles of the pelvic limb of the llama' and other such articles, we rejected as being too technical and probably unhelpful for our purpose.

I looked forward eagerly to the arrival of the articles. This sort of research fascinated me and it held the charm of an exploration. I was still brooding about writing the book and I hoped that extra knowledge would give authenticity and interest to any writing that I did, besides helping us in our llama management.

# 17 Mr Widdle

The spring of 1976 was here. The larches in the forest were green, and a delicate tracery of new leaves was appearing on the oaks and beeches. The carpets of early daffodils at Tŷ Mawr were nearly over, and the sunshine was warmer and the days longer.

Apart from a little frost, the past winter had been mild and, on the whole, our livestock had fared well. Our two pretty Welsh Mountain Pony mares, Carys and Wyspa, each had a colt foal. They had been born within three days of each other, and now played and ran races together in the spring sunshine. Star, our black Welsh Riding Pony, had no foal this year. We didn't mind. She had bred some lovely foals and she deserved a rest now and again. Arm's best pony, Dolmen, was Star's son, by our own stallion, Idris. Ann loved Dolly dearly. She had broken him in herself, struggled fiercely with his stubborn temperament, soothed his fears and taught him to jump when he thought he couldn't. Sometimes she rode him sitting backwards in the saddle, and practised circus tricks—adding a few more grey hairs to my head. Ann loved him and bossed him and they had a marvellous relationship.

Dolly was six years old now. He was a liver chestnut with three white socks and an off-centre blaze down his face. He was a sturdy pony, just about 13 hh and he suited Ann well, though he didn't have quite enough shoulder to be the ideal riding pony. If Ann's horsemanship continued to improve at its present rapid rate, she would need a better mount before long. It was doubtful if we could afford to buy one for her. We should have to breed it, and so we planned to put Star to an Arab stallion to see what we would get.

Ann was now ten. In the autumn before, her Grandma had paid for her to have five riding lessons at a superior riding school but, apart from that, she was self-taught. She was long past the stage where I could help her, except in a general way. She could have had more lessons if she had wanted them but Ann said she didn't. She felt that the riding school ponies did not enjoy themselves and so she could not enjoy riding them. She missed the zest and enthusiasm of her own darling pony. Now she picked up the theory of horsemanship from books and seemed to put it into practice with good effect.

In the autumn, she had entered a small, local cross-country race, riding Dolmen, and had come second in her age group, winning a blue rosette and a mane comb. Her appetite for competition was whetted.

'When can I ride in a proper gymkhana, Mummy?' she kept asking. 'I've been reading about them. I do want to ride in one.'

So now Ann waited eagerly for the summer, in the hopes of going to gymkhanas. She practised busily with Dolly, and painted show jumps with the help of Amanda and Penny and, occasionally, John. Our fields were littered with extraordinary obstacles, and ponies and children were frequently to be seen flying over them. Riding was Ann's passion and delight now, as it would have been mine if I had had the same opportunities at the same age. Paul and I felt that we must give as much support to her as we could afford. I only wished that I was more knowledgeable about the highly specialized, esoteric world of competitive riding—and I was sadly aware that it was really a sport for the rich.

Now the lambing was almost over and we seemed to have quite a good crop of lambs. In good weather, the lambing of a mountain flock is a fairly simple affair. The Welsh Mountain ewes are excellent mothers and, in most cases, are best left to themselves to get on with the job of giving birth and bringing up their offspring. There are just a few cases of mispresentation which need help, a job I didn't like and was not good at. Sometimes there is a young ewe which lambs without much milk and which tends to keep forgetting about its baby until the lamb falls prey to a fox or carrion crow. Our flock is spread out over such a wide area of rough ground that a lot of walking is needed just to keep an eye on it. It is bad weather that brings the trouble but we had been fortunate that spring.

Mooey was a marvellous mother and again she had twins. Brown Patch, another of the children's pets, had lambed for the first time, producing a minute little lamb, which Beenie named Dinky. Brown Patch was one of those ewes which doesn't bother much about its lamb, and little Dinky was always getting mislaid. His thin bleat was often heard, as he stood miserably stranded with not a sheep in sight. His mother was probably away attending the hen feeding, where there might be a little corn to be had, or else going round the ponies' feeding bowls. After a few days, he realized that he had to run to keep up with his mother's rapid progress round all the likely sources of extra food, and he raced after her like a frail woolly spider. Brown Patch was both docile and greedy, and Ann soon found that the ewe could be tempted round a small jumping course for the sake of a few nuts. When Dolly and Ann had finished their jumping, it was Brown Patch's turn to go over the hurdles. She proved to be both willing and athletic.

One day I came on a young man, dressed in climbing gear, standing by the cowshed. He had a young lamb in his arms. He told me that he had found it trapped between some rocks on the mountain and that its mother was nowhere to be seen. He said that he had been climbing on Christmas Buttress, the rock climb above the Clogwyn pens, climbed and named by Paul years before. I thanked the young man and took the lamb from him. I was not too pleased. Generally, if someone finds a lamb which seems deserted, it is better to leave it where it is and to tell the farmer so that he may go and see for himself. The chances are that the lamb is not really deserted and that the mother will come back to it when it has been crying for a while. If the lamb is taken away, the ewe will be upset when she returns and may travel quite a distance looking for it. Now I was faced with the job of taking this lamb back to the mountain again and trying to find its mother.

After lunch, I set off up the mountain with the lamb tucked under one arm. Beenie and the llama came for the walk too. High up among the rocks, the signs of spring were less obvious. Last year's dead grass and the tough old stalks of heather still had a bleak and wintry look, though, here and there, the tiny yellow flowers of tormentil starred the turf. Turning southwards, we could see the springlike greens of the valley stretching down to the sea, but to the north, as we climbed upwards, the land was steep and bleak. There were not many sheep up here and we could see none that was a likely mother for the lamb.

Ñusta had never been so high with us before and now hurried along enthusiastically, picking her own route and occasionally pausing to graze the new herbage. She seemed to enjoy the wide vistas and would gaze keenly around her as she reached each fresh vantage point.

We reached Christmas Buttress and the Clogwyn mountain pens. It was right up here, under the sky and the grim, grey crags, with a great view of mountains on every side, that the Clogwyn flock had been shorn in years gone by. Then, the farmers had found it easier to carry the fleeces to the lower barn than to drive the sheep down with the wool still on their backs. Perhaps the weight of wool was not so great in those days, and many of the small enclosures below, now shoulder-high in bracken, had been the farmer's precious hay fields, where a trampling flock could do much damage. That was long ago and only the ghost of the great activity, which had taken place there once a year, was left. Now the pens were deserted, with clumps of rushes growing within their walls and a few stones fallen here and there.

From the description, this was the spot where the climber had found the lamb. Now we must get it to bleat, and its mother, if she were anywhere about, should show up. I put the lamb down.

'Let's nip quickly through the gateway before he follows us,' I said to Beenie.

We nipped quickly into the pen while the lamb was looking the other way. Although he was so small, frail and young, the lamb had his wits about him. He nipped after us with great celerity.

'Run, Beenie,' I cried.

We dashed through the gateway on the other side of the pen and ran as fast as we could up the Heather Hills of Clogwyn Hafodty. When we slowed down, gasping, the lamb was right behind us. Having been deserted once, he seemed determined not to let it happen again. We had a problem. So long as the lamb followed us he did not bleat—but bleat he must, if his mother were to be found. We solved the difficulty in the end. Beenie retreated over a near-at-hand skyline, while I held the lamb. Then I put him down in a dense, knee-high patch of heather and, with big, leaping strides, dashed away through the long growth, scrambled up the rocky shelf ahead and ducked down. Presently I ventured a look. The lamb was discouraged now and was having difficulty in hopping through the rough stalks in our direction. Beenie and I retreated rapidly to higher ground, where we were too far away to attract the lamb but where we could see what was happening. We felt very mean as we sat on our high look-out and observed the little white dot far below struggling through the heather, its feeble wail coming to us on the still afternoon air.

Ñusta had also followed us through the sheep pens and had begun to graze some distance away. The dry heather was very much her sort of food and she was making the most of it, at the same time keeping her radar scanners tuned to Beenie and me.

There were no sheep in that particular place and no maternal bleats answered the lamb's cries. There were a few ewes further away on the Heather Hills so, leaving Beenie to keep watch, I walked on to take a closer look at them. Most of them had lambs of their own. There were two without; one looked definitely barren, old and rather spinsterish, as though she had not had a lamb that year and was not going to have one; the other I wasn't sure about. She looked quite young but she didn't react in any way to the sad crying of the orphaned lamb. I walked round the group of ewes and drifted them a little towards the lamb. None of them took any interest in its cries. Like all the Clogwyn sheep, they were a wild lot. When they saw Beenie lower down and realized that I was trying to get them to do something, they became alert and nervy. The mothers called their offspring. They began to look round for the best way to disappear. Then, with escape from interference uppermost in their minds, they melted rapidly away in the dips and hollows of the rough ground. I explored a bit farther, looking round for any dead sheep which might account for the orphan, but there were none. So I returned to Beenie.

'None of those will make a mother for it,' I said, 'and I can't find a dead one.'

We waited a while longer, hoping that a ewe would turn up from somewhere, but none did. We scanned the distant specks of grazing sheep at the far end of Hafodty, but they all seemed quite settled. None looked as if it were searching for a lamb.

'Ah, well,' I said. 'Let's take him home. I don't think we're having much luck here.'

Beeenie agreed. Neither of us liked listening to the pathetic wails and we should be much happier when we had given the orphan a bottle and made him comfortable. We had tried our best to find his rightful mother but without success. Now we must take her place.

We set off down the mountain, carrying the lamb. Ñusta followed. Although the lamb was so small, he seemed to grow heavier as time went on. We set him down to walk but he would go for only a few yards before collapsing. He was getting weak from hunger and all the crying. We carried him in turns till at last we reached Tŷ Mawr again.

Some new lambs are unwilling to take a rubber teat for their first few feeds. You have to stuff the teat into their mouths and sometimes use an old-fashioned, open-ended bottle. You can then blow down the open end and trickle some milk down their throats if they won't suck. This lamb had no sucking problems. When I presented him with a bottle of warm milk and glucose, he latched on to the teat avidly and did not stop sucking until all the milk was gone. You could see him reviving.

When they came home from school, Ann and John were delighted to find that there was a pet lamb again.

'What shall we call him?' I asked.

'Snowy.'

'Snow White.'

'Lambsy.'

'Woolly.'

Unfortunately none of these nice names seemed to stick. He was an exceptionally bright and intelligent little lamb, but he ended up by being called Mr Widdle. He would sit or wander about outside the cottage for hours but, the moment he gained entrance, he would hurry to the mat and make a puddle or, worse still, jump on to the sofa and do it there. Apart from that, he was a charming little animal.

He couldn't be allowed indoors much, but we all loved him dearly.

When he was a couple of weeks old, we noticed that he was beginning to take up with the llama. Most pet lambs are human-orientated. They potter about outside the house until the front door is opened and then rush up demanding a bottle. If you go for a walk, they follow you just as they would their mothers. But Mr Widdle didn't; he followed the llama. Although no milk was forthcoming in that direction, those long white legs seemed to attract him and to give him his bearings. If he wanted food, he came to us, but for company he went to the Um.

Paul was amused at the sight of them together.

'The llama and the llamb,' he said.

They certainly made an odd pair, Ñusta walking ahead with Mr Widdle a yard or so behind. They almost matched, but not quite; that was the odd part of their appearance. Ñusta was either a very elongated mother or Mr Widdle was a sawn-off child. You couldn't decide which of them was wrong.

The relationship was one-sided. After an initial show of kind interest, Ñusta decided that she definitely did not want a foster child. Over and over again, she made it very clear that Mr Widdle could push off, but he never took the hint. She would set off to go somewhere and there was Mr Widdle right behind. Where Ñusta grazed, there grazed Mr Widdle. When she lay down, he lay down too.

He liked to lie against her fluffy side, as lambs do with their mothers, but Ñusta thought it was insufferable cheek. She would snake her head at him in a threatening manner but he seldom moved. It was usually she who had to get up again and find another sitting place. Once or twice, we saw him jump on her back, when she was lying down, and begin to hop about. This is a normal game for a young lamb and its mother will just sit there with a complacent expression on her face. Not so Ñusta; she refused to consider herself as Mr Widdle's mother and objected strongly. Mr Widdle, playing King of the Castle, would be confronted by a furious llama's face, spitting violently, and then his fluffy castle would heave itself up, as Ñusta struggled to her feet, and he would be tumbled to the ground. This never stopped him from trying again.

Sometimes they would sit together in apparent peace. I remember walking down from Carneddi one day in late spring. For some reason, Ñusta and Mr Widdle had not walked up with me and now I was looking about for them. The view was beautiful on every side, with the fresh green of the new leaves. A cuckoo was calling from the woods. Paul had recently ploughed the patch of ground where we grew our potatoes, a few swedes and other root vegetables, and a little kale for the cows. The ploughing lay below me now, a neat rectangle of freshly turned soil, all ready to be planted. Then I saw the missing pair. In the very middle of the patch of bare earth lay the little figure of one llama and, at a respectful distance of about four feet, a white speck which I knew was Mr Widdle. It was strange to look down on those two animals, sitting side by side in the middle of the ploughland, a mini-desert for them with not a plant within reach. Ñusta, we knew, was attracted by dusty ground—perhaps it was reminiscent of the Andes—but it was unusual for a lamb to choose such a spot. Anyway, there they were, together but not contiguous, maintaining a polite gap of a few feet between them. They were both facing the same way and seemed to be gazing thoughtfully into the distance. I wondered what, if any, communion passed between them.

As Mr Widdle grew bigger, he became more and more cheeky. He would butt Ñusta's legs and try to play ram lamb games with her, though he was too small to reach higher than her hocks. She did everything she could to show him that these attentions were unwelcome, but he thought it was all part of the fun. If she couldn't escape from him, she would go into a tantrum of spitting and head-snaking but it was no use; Mr Widdle came back for more. Once we saw Ñusta, driven to complete desperation, resort to her final aggressive act. When the spitting and snaking had proved useless and Mr Widdle was still in great form for more rough play, she knelt on him, tried to obliterate him and blot him out with her broad chest. We quickly rescued the lamb from underneath her. He was none the worse and had only got what he deserved, but I began to feel that his

constant pestering might sour the llama's normally courteous and amiable behaviour.

Mr Widdle's jolly and extrovert character was attractive
until it became a worry and he began to carry his games
too far. Brenda had a large, square, red handbag which Mr
Widdle liked. When she brought Helen up to tea in the
afternoons, Mr Widdle would be on the look-out for that
bag. He would put his little head down with its sprouting
horns, charge and slam into the bag.

'Olé!' cried Brenda.

Ann, John and Helen were in fits of laughter at this
display, but it became less funny when Mr Widdle took to
butting small children. John, a sturdy seven-year-old, was
more than a match for him, but the lamb grew to be quite
terrifying to a dainty four-year-old like Helen.

There was only one thing for it—Mr Widdle must go. In
the following autumn, very regretfully, we included him in
a lot of store lambs for the sale. The menace of Mr Widdle's
presence was removed from the lives of the llama, Helen and
other small visitors to the farm. We were sorry to lose a lamb
with so much character; we should not soon forget him.

# 18 Llama Revealed

The month of May came. It was a busy month but, in spite of all the farm work, I now made a serious attempt to write the next book. For a couple of hours on most days, I would retire to the Writing Hut below Tŷ Mawr cottage. Here, years ago, I had written part of another book. The hut had once been a henhouse but, with the perches and nest boxes removed and the wall lined with hardboard and the floor with linoleum, it made a good and peaceful retreat for me. It was screened from the cottage by a dry-stone wall and overhung by a huge ash tree. Round about, clumps of bracken were beginning to unfurl and foxgloves were pushing up their flower spikes.

It was a secluded spot. If I left the door open, which I usually did when the weather was fine—and it was very fine that May and June—I could watch the birds in the branches of the ash tree. It was a beautiful scene, reminding me of an exquisite oil painting framed by the doorway of the hut. The massive trunk of the ash, rough and greyish, rose from behind the moss- and lichen-covered stones of the wall. About twelve feet from the ground, there was a niche in the trunk where a flourishing colony of hart's tongue ferns grew, brilliant green and trailing against the grey bark in a kind of studied elegance. The foreground was veiled by a lacework of fresh leaves. I could watch the birds without disturbing them, perhaps a little hedge sparrow, a tit or two, a wren and now and again the noisy arrival of a jay.

At first, the llama was surprised to find me in the Writing Hut. When I disappeared in that direction, she galumphed after me and gave a high-pitched toot when she found me within. She seemed to like having me visible and within reach, and stayed close to the hut so long as I was there. She had a chew at the ancient linoleum by the open door and I had to tell her to stop it. Then she had a chew at the rotten woodwork round the window and I had to tell her to stop that too. I didn't think it would hurt her but it would certainly hasten the disintegration of the old hut. Then she put her head through the windowless window and snatched half a page out of my loose-leaf book.

'Are you censoring it?' I asked her as I moved the book out of range.

'*Mmm*,' said the llama as the last corner of the page disappeared into her mouth. Then she craned in again to try to get another sheet.

'NO,' I said firmly. 'Hop it.'

Since it seemed to be NO to everything, the llama pottered off to graze the heather on the rocks nearby. I worked on, but every now and then I would raise my eyes to look at the green glade outside and to watch the wild birds, a few hens, our pigeon and the llama going happily about their business all around me.

As midsummer approached, we began to think again about shearing Ñusta. We were still not sure whether it was the right thing to do and felt we needed more information. We had heard that the alpacas at Chester Zoo were sheared every second year because they did not breed regularly unless this were done. I didn't quite see the connection but it seemed to work. Now there was a flourishing herd of alpacas at the Zoo, all bred from a small nucleus. We were told that a farmer from North Wales did the shearing and had the fibre in payment for his work.

Shearing an alpaca, we thought, might not be so difficult. They were no bigger than a long-legged sheep—but a llama would be a different matter. A full-grown animal, we read, might weigh as much as 350 1b. We judged Ñusta to weigh something under 200 lb, but even so it would be impossible to handle her according to the Bowen method of machine-shearing which we normally used for the sheep. We thought of throwing her down and tying her legs to restrain her, but this seemed an unkind procedure with such an intelligent and sensitive animal. It would be an affront to her dignity and a gross betrayal of trust. We rejected the idea as soon as we thought of it. Then Paul wondered if we could administer a tranquillizer, but it seemed ridiculous to have to resort to drugs for such an ordinary task as shearing. Perhaps the best method for collecting the fibre was by combing after all.

But we were still not satisfied that we had enough

information. Beenie wrote with queries to some of her friends who had been in Peru. They replied that the Indians sometimes shampooed their llamas and 'clipped them like sheep', adding that care was taken to shear only in warm weather in case colic should result. This was most interesting but, after the swimming episode, I did not feel enthusiastic about shampooing. Also I didn't see how llamas could be 'clipped like sheep'. Clipped, yes, but not like sheep; llamas were so much bigger. If only, I wished, I could make contact with a real, genuine llama farmer to find out the practical details—but I did not know where to begin.

Then, the owner of a holiday cottage in the district, knowing of my interest in llamas, sent a photograph which she had taken during her travels in Peru. She had taken the picture as a colour study, and a good one it was too. It showed a group of four or five llamas together in a pen paved with yellowish slabs. The llamas were white or light brown, with one unusual animal in their midst, a dark grey with big black spots on its body. The particular interest to us was that the llamas were obviously newly shorn. Their elongated necks looked longer than ever and their svelte, greyhound-shaped bodies were revealed. You could see the clip-marks of hand shears clearly on them. Here was proof positive that Indians did shear llamas.

But we still couldn't quite make up our minds.

Brenda and Helen were staying at Gwylfa.

'You mustn't shear the Um,' said Brenda. 'It'll ruin her appearance and I won't allow it!'

Beenie thought so too, also that the fibre would begin to rise naturally, like the sheep's wool, and that we should be able to brush off larger quantities as time went by. Indeed, when I parted the fibre along her spine, I could see signs of a slight rise, though it didn't seem to be coming off any more easily.

Paul was convinced that we ought to go ahead and shear Ñusta. He was pleased by the sample that Mrs Arnett had spun for us and wanted to have more. My feelings were rather mixed. I wanted to have enough fibre to make jerseys for the family but I didn't like the idea of having to man-handle the llama to get it. I wanted her to be more than just an exotic pet but I also wanted to be absolutely sure that we were doing the right thing.

Then someone sent us a newspaper cutting. It seemed that the Peruvian Government had stopped exports of llama and alpaca fibre in an attempt to encourage home manufacture. Because of this, the manager of a famous woollen mill in Yorkshire, which specialized in the spinning of mohair and camel, llama and alpaca fibre, was planning to breed llamas and alpacas in Yorkshire to supply the firm's own raw materials. The first two animals had arrived, a male and female llama from Peru. The shearing was going to be done shortly.

This was very interesting. I wanted to know more. I wrote to the manager, Mr Bell, enclosing a stamped, addressed envelope, and asking him if he would be kind enough to let us know how he was going to do the shearing. I received a friendly reply by return of post. Mr Bell had not sheared his llamas yet but, when he did, he would let us know how the proceedings went. His llamas were large adults and he had been advised from Peru to tie their legs for the shearing. He was interested to hear that we had a llama too; he did not know that there were others in the country, kept in farm conditions.

Paul continued to be enthusiastic about the idea of shearing Ñusta. He began to practise simulated shearing on her to get her used to the feel of it. She was always in a sunny and good-tempered mood first thing in the morning when he let her out of the stable. After a night on her own, she was delighted to see someone again and would toot and breathe into his face with pleasure. Now, as part of the morning grooming, he passed a monkey-wrench through her fur for a minute or two, clicking it as he did so. Later he used a pair of hand-shears, pushing the handle through the fibre and snipping the blades in the air. She didn't mind in the least, he reported. He thought that if we both set to work on her in the morning when she was cheerful and relaxed, we might shear her standing up without upsetting her. I was willing to have a go, but somehow we kept putting it off.

In the middle of June the weather became very hot. We had already had a warm and beautiful spring. Now, for a few days, we were enveloped in a sea mist, white and luminous, and unusual here. Then the mist thinned, the sun came out and the temperature climbed to the nineties. I had not known it so hot before. Sometimes there was a breeze, but it was like the warm breath from an oven and brought no coolness with it. The nights seemed almost as hot and the slightest exertion made us stream with perspiration. Even just sitting, you could feel it streaming down your face. The hens no longer sunbathed luxuriously in their dust-baths but loitered in the shade with their wings held away from their sides to let the heat out. The ponies sweated and swished and stood head to tail in the deepest shade. The cattle buried themselves in the bracken to hide from flies or emerged, frantically galloping, tails over their backs, their black coats sparkling in the glare of the sun, to escape from their tormentors. The heavily clad sheepdogs became subdued and flopped down, panting, on any cold stone floor where people would trip over them. The mountainside seemed to be bare of sheep. They lay hidden in the bracken or in the shade of a stone wall, panting and enduring under their heavy fleeces, until evening should bring them some relief and enough energy to graze again. The ground baked and shimmered in the heat. The stones reflected the glare from the sky.

It was the hottest June on record, we were told. The temperatures in some part of the British Isles were higher than those in Rome and Madrid. On most days, the children went to bathe in the pool at Tanrhiw after school; it was the only way that they could get cool but they were hot again before reaching home.

Most of the country was suffering from a severe drought after the dry spring, but in North Wales we were lucky. There had been enough rain to keep the mountain bogland saturated and most of the grass green. Water still ran merrily into our storage tank but, over the rocks where the soil was thin, the grass was beginning to yellow. Evaporation must be high and we hoped that more rain would not be long in coming or we should be in trouble.

Ñusta did not seem to be affected by the heat. She would sit out in the blazing sun, with her lashes a little lowered and the pupils of her eyes reduced to slits by the bright light. She did not pant or seek the shade but remained quite unruffled. This summer, I did not see her adopt the strange attitude she had sometimes used in the hot weather of the year before, when she had crouched with her hocks upright behind her to allow the air to pass under her belly. She did, however, seem to like to lie in the sitting-room, where the thick walls of the cottage gave some coolness. She would loll there, half over on her side, and, when I laid my cheek against the top of her little head, it was bedewed with sweat. Her fleece was much thicker than last summer. Perhaps she would be more comfortable without it.

The hot weather made it important to get the shearing done without delay. The sheep were suffering in their thick fleeces and the danger of fly-strike was increased.

We planned to get extra help with the shearing this year, hoping to make time for me to go on writing the book. In our early years at Carneddi, shearing had been a communal event. The neighbouring farmers had banded together to wash, shear and dip each man's flock and to earmark and castrate his lambs. These sheep days had been pleasant social occasions on which a great deal of work had been accomplished. The farmers' wives and daughters provided refreshments for the visiting shearers. Countless plates of thin bread and butter, salads and tinned meat or home-killed mutton, rice puddings and rhubarb tarts were consumed. Countless cups of tea were drunk, and chilled butter-milk was brought out as refreshment to the sheep pens, where six or eight neighbours would be sitting on their benches, clipping away with hand-shears. The work was all done by nightfall and that was the end of it. Old John Williams of Beudy Newydd, our neighbouring farm, seemed to fix the order of the shearings in our district. Nobody began haymaking until all the shearings were done. Then, with John Williams's death and the advent of machine shearing, the old order began to fall apart. There were fewer farms. The farmhouses became holiday cottages and the land rough grazing. The original people died or moved away and mostly the younger generation did not want to take their places. If a man helped his neighbour now, it was usually for money. We regretted the passing of the old community.

For several years, Paul and I had been managing our shearing piecemeal between us, with some outside help for the gathering of Clogwyn. Here the land was a wilderness of hills and valleys, trees and bracken, and no two people could make a clean gather of it by themselves, however good their dogs, and, after Ruff and Del died, our dogs were not so good. The shearing had dragged on for days sometimes and, though we still liked doing it, the job had lost a little of its old charm.

This year, Arthur promised to help us with the work. We had known him for the thirty-one years we had been at Carneddi. His home had been the little farm of Cwm Caeth behind Nantmor village and he used to help us at the old communal shearings. Then, after his father died and the farm was sold, he moved away to work on another farm a few miles distant. Later he took a job with the County Council but spent his holidays and many weekends working with sheep round the district. He was a wonderful shepherd. It was a pleasure to watch him with his dog and to hear the versatile language of his whistles, interspersed at times with a high-pitched shout of command. He was a tall man with powerful, stooping shoulders, a mass of reddish hair and a fast, loping stride over the hills. He was well-endowed with that special Welsh sense of humour, which has a quick and funny answer to most questions. I had been threatening him for some time that it might be his job to shear the llama, but he just laughed.

We gathered the Clogwyn sheep very early one morning before the full heat of the day. Later it would have been impossible to move them out of the shade. Then Arthur, his son and Paul sheared them by machine, while I did a few by hand. I never really liked machine-shearing. It always seemed tremendously hard work, probably because I had not mastered the art entirely. The machine chattered away frantically, with never a moment to pause and rest your back. I could do it after a fashion, but if there was anyone else there to use the machine, I let them.

In the days of hand-shearing, we had worked in Arlas, the small field adjoining our sheep-pens. With the advent of a machine, we needed to be near the electricity supply. Now, each year, we arranged a shearing area in the carriage shed, a building which Paul and I had erected years before as a garage and store shed. It was very convenient. We had a pen to hold fifty or sixty sheep on a slatted floor in one bay, room for two machines and a hand-shearer or two in the next, and a tarpaulin slung up to hold the wool in the third. We rigged a race from Arlas to the carriage shed so that we could drive the sheep up and down without difficulty. If we could pen the sheep dry, we could go on working even if the rain came.

Now, even in the shade of the building, it was unbearably hot. As I sheared, a continuous stream of perspiration dripped off my nose and chin on to my sheep. My glasses were steamed up and I don't think I had ever been hotter.

It was even worse for the machine-shearers and for J., who was here for a fortnight and who was helping with the catching and the fleece rolling.

The llama, of course, arrived to see what was going on.

'Your turn next,' we told her, but in fact we had too much work that day to think of the problems of how to shear a llama. She peered through the wire mesh of the carriage shed gate, taking in everything that was happening with a long intent stare. Then she turned away, folded her long legs and settled into a sitting position on the dust and stones of the yard, next to a couple of sheepdogs. They were crouching in a strip of shade by the building, panting furiously and never for a moment taking their eyes from the sheep inside. The llama sat in the dazzling sunshine, observing all the activity with great interest.

When the first batch of sheep was shorn, Paul and Arthur drove them up the race and round the cowshed to the pens. We had taken a couple of hurdles from the race to use for something else and now there was a gap in the defences. J. and I stood in it to turn the sheep. The llama got up and came to stand beside us. Then we had to wait until the next bunch of woollies were driven down for shearing. It was a critical point. The outgoing sheep would run up willingly towards the pens but those coming down would be afraid to enter the building and would break out if they could. We waited alertly for their coming. Away from the shade, the sun beat down on our heads, like the heat from a furnace, and reflected back from the baked stones of the yard. Every light surface dazzled in the glare and even the green grass glittered in the sunshine. It was stunningly hot.

There seemed to be some hold-up at the pens. We went on standing in the sun as the minutes passed. The llama was tired and sat down again, but she sat just halfway between J. and me, in line with us and facing into the race. It looked as though she had taken up this position on purpose. We wondered what she would do when the sheep came rushing down the race. Then we could hear them coming. Then they turned the corner and were upon us. J. and I were ready to repel them if necessary. The Um seemed ready to do her job too. She did not get up but, as the sheep came flocking past, hesitated at the sight of the gateway into the building and were half-prepared to turn back and try to break out past us, she stretched out her neck towards them, snaked her head and pulled nasty faces. At the sight of J., me and the cross-looking llama filling the gap in a menacing line, the sheep had second thoughts and pattered on to the slats in the shed in a bunch, while Paul rushed up behind and closed the gate on them.

'Well-trained sheep llama!' I said to J.

He could only agree. The llama's behaviour had appeared uncannily as though she understood just what we were trying to do and wanted to help us as much as she could.

The delay at the pens, we learnt, had been caused by the death of a sheep. One of the newly shorn animals had been seized by convulsions and had died in a few seconds. We could only conclude that she had been affected by heat stroke.

Two days later we were ready to shear the Carneddi flock. The weather was still just as hot, and again the gatherers went up the mountain early before the full heat of the day made the sheep unwilling to move. There were fewer sheep in this flock and only Arthur came to help us. By mid-afternoon, the shearing was done and, thankfully, we collapsed into deck-chairs in the shade of Carneddi garden, drank tea and recovered from the effects of the heat and the work.

'What about doing the llama now?' I said.

All of a sudden it seemed a good time to tackle the task. It seemed a pity to leave Ñusta sweltering in her fluffy robes while all the sheep were shorn, cool and athletic once more in their underwear. Arthur was available and, being an excellent shearer, might make a better job than Paul or I. We had not yet heard from Mr Bell, but his experiences of shearing llamas might not help us a great deal since we did not plan to tie Ñusta's legs.

'I'll have a go,' said Arthur. 'Better use hand-shears. We might go too close with the machine.'

Paul also thought it was a good time. He, Arthur, J. and Becky would dip the newly shorn sheep while I prepared for llama shearing. The last cup of tea drunk, we hurried off to our various tasks. I went down to Tŷ Mawr to collect the llama's brush, collar and something to put her fleece in. This was going to be an historic moment. She came galumphing down over the rocks with me, followed by Mr Widdle, and entered the cottage. I put a handful of pony cubes in her bowl in the kitchen.

'Have a few to keep your strength up, Tootie,' I said. 'You are going to be shorn.'

I found her brush and collar. The lead appeared to be missing but perhaps we wouldn't need it. What could I use to put the fleece in? We were short of clean sacks and I didn't want to put the fibre in something which might sully it. I grabbed a couple of clean pillow-cases and then looked round for some acceptable reward. In this heat, Ñusta's favourite Maltesers would be melting and uneatable so I found a packet of Maryland Cookies in the tin. That seemed to comprise all the extras that would be needed for shearing a llama.

Then we walked up to Carneddi again, Ñusta, with her satellite Mr Widdle, following at a stately pace a few yards behind.

The dipping was still in progress, so I set to work to clean up the shearing area in the carriage shed. I excluded Mr Widdle but Ñusta came in with me. She examined everything with close attention and watched as I picked up all the snippets of sheep's wool and gave the concrete floor a thorough brushing. Next I brushed her, removing as much of the dust, dirt and hayseeds from her tresses as I could. Then I swept the floor again. The llama seemed relaxed and happy, rather enjoying the shade of the building and the interest of being inside it instead of only looking through the wire gate.

Presently Paul and the others came down from Arias, the dipping finished. Now we would begin. There was a slight pause while Paul and J. organized their cameras. The light was getting poor in the carriage shed now, as the sun lowered towards Moel Hebog and a bank of haze rose to meet it, but photographs we must have, whatever the light. Ann and John, home from school now, were longing to see the operation begin. Beenie went to the house to fetch my mother, who must be present at such an interesting moment. After all, if it weren't for her, there would be no llama. She sat down on an old egg crate in the corner, where she could see everything. Arthur was sharpening the best pair of hand-shears. The llama watched us all with interest.

She rather liked Arthur. He had a relaxed and confident way with animals. When he approached her with the shears, her ears went forward and she was prepared to talk to him but, when he stooped and began to snip into the fibre of her left shoulder, she turned huffily away with her ears back. I offered her a Maryland Cookie, but she was not prepared to be bribed.

'I'll put the collar on her,' I said.

'Yes, I can do her if she stays still,' said Arthur.

I buckled the collar round her neck, took a firm grip on it and wondered what would happen next.

'Stand still, Tootie,' I soothed her. 'We are just going to take your dress off.'

Ann offered the cookie again but the llama wasn't interested. She surged forward as Arthur came close and began to snip again. I withstood the pressure and Arthur continued to shear. Then she pulled back hard, giving a harsh little toot of distress. I clung on firmly, while Arthur's powerful frame was unmoved and he continued to work away. Then she came forward again but still I clung on. I was aware of Paul and J. to one side, taking photographs or ready to help me if the need arose. The children were already picking up fragments of fleece, which was now coming off like a curtain, to go into the clean pillow-case. Then, quite suddenly, Ñusta gave up and stopped struggling. 'They are but feeble animals,' Goldsmith had written. Her struggles had not been very violent and I had been able to stand my ground. Now she had given in. She stood stock-still, her head and neck very upright, her ears out sideways. I could feel the tenseness of her body, but outrage had now given way to resignation. Arthur worked fast, clipping carefully and quickly, and the fleece was coming off like a big ragged shawl. As he began to shear up her ticklish left thigh, the llama slowly wilted down into a sitting position, her hind legs collapsing first. We could almost hear her saying: 'What can't be cured must be endured.'

'All the better,' said Arthur, squatting down and continuing with his work. Now he had come to her tail, the beautiful feather duster.

'Shall I do this?' he asked.

Paul said: 'Yes.'

With a few snips of the shears, the long, gingery locks fell away and, instead of the feather duster, a thick, muscular, ginger pipe-cleaner, about twelve inches long, was revealed. Then Arthur was ready for the right side. The llama was now sitting close to the back wall and I drew her towards the left so that he had room to work. She did not resist or try to get up. I found that I was still clutching the collar very tightly and I released my grip. Ñusta straightened her neck to a more comfortable position but did not attempt to rise. I took the collar off but still she sat, not happily but with a determined dignity that I found touching. I had been tense in sympathy with her but now I began to feel happy that the shearing was going so well and was nearly done. It was simpler than we could have expected. I began to feel like a mother whose child has behaved unexpectedly well in difficult circumstances.

Then she was finished. Arthur stepped back to look at his work. The children gathered up the main fleece and packed it in a pillow-case. Ñusta went on sitting, very alert and a little tense, with an 'is it all over?' expression on her face.

'Come on, Tootie, you can get up now.'

At my words, she sprang to her feet and stood there for a moment, no doubt surprised at the feeling of her fleeceless state. Then this new, nude llama walked towards me in her old relaxed manner. The shearing was an incident that she was now prepared to forgive and forget. Arthur trimmed a few ragged pieces on the flanks which he had been unable to reach whilst she was lying, but she did not object.

I thought she deserved a reward.

'Let's try her with a biscuit now,' I said—but the on-lookers had eaten the lot.

Arthur had made a very neat job of the shearing. It had been a bit like unwrapping a Christmas parcel; we didn't know quite what we would find inside. Now we gazed at the llama revealed, her contours no longer hidden in a cloud of draperies. She stood there, a svelte, greyhound-shaped animal, the short, plushy fibre stippled with the marks of the shears, the long neck seeming longer, the extraordinary tail poised in a perfect question mark.

'We haven't ruined her,' said Paul. 'She looks nice.'

Everyone agreed with him. We thought even Brenda would not mind the transformation. Ñusta looked nice but she looked different. We couldn't quite get used to the new outline, particularly of the tail. It was the most un-usual tail we had ever seen. It fitted neatly between her pel-vic bones, like the lid of a box when she carried it down, which she seldom did. For walking, we noticed, it assumed a backward-facing question-mark position, held high above her back, like a small, inadequate counterbalance to the long neck at the other end of her body. If she met a friend or was offered a Malteser, the tail went slowly upright and then curled up to make a small cup-handle over her back. When grazing and if there were flies about, she would hold it away from her body and curved downwards, whisking it briskly from side to side. It was strong and mobile throughout its length, and she could move just the tip of it like a cat. When she sat, the tail was just the right length to touch the ground, where it rested like the end of a huge finger.

That evening we weighed our harvest of llama fibre. There was a little over 3½ lb. It was, perhaps, less than we had expected but there was more than a pillow-case full. It seemed strange to see a quantity of llama fibre now no longer connected to the llama. Arthur had left her with a covering of about three-quarters of an inch so we thought she would be unlikely to get a chill or colic.

A few days later a friend, Professor Allen, called to see us. He was interested in the newly shorn llama and amused by her extraordinary tail. I mentioned we had been wondering what the tail position actually was, before the long, gingery hair had been clipped from it.

'You could have guessed that before you sheared her,' he said.

'How could we?' I asked.

'It's a well-known fact,' said Professor Allen, 'that most pieces of research end in a question mark.'

# 19 A Hot Summer

It took us a little time to get used to the new, streamlined appearance of the llama. The children could now no longer call her FluffybUM. She had lost something of her prettiness with the loss of her fleece but still she looked trim, elegant and unusual. It was possible to get a better idea of her anatomy now that it was not swathed in fluff. The little, tucked-up wasp-waist was in greater contrast to the deep chest and rib-cage which housed her super, specially adapted, high-altitude lungs. You couldn't feel any suggestion of shoulder blades at her withers; they seemed to be hidden under the attachment of her powerful neck muscles. You couldn't easily feel her hip bones either. The activities of her newly revealed tail continued to fascinate us.

Now, with the summer weather so fine and dry, the Um slept out every night. In the mornings, we would usually find her seated on the slates outside the cottage, waiting quietly until the family woke up and invited her indoors for a feed. During the day she was busy about the farm, taking a llama's eye-view of all the activities and joining in whenever she could. This llama's eye-view included looking out for my return if I had been away in the car. She was aware of the sound of the engine before anyone else could hear it. As I drove up the hill, I could usually see a small white face with dark eyes and long ears peeping inquiringly over the five-foot wall which ran along the crest of the spur. After a long look, the face would disappear and I knew that the Um was rushing along on the other side of the wall, to reach the yard gate at the same time that I did. When I stopped by the garage, a head and a very long neck would come in through the car window and the llama would toot an excited welcome. I usually kept some Peanut Treets in the glove compartment to reward the faithful animal.

In the evenings, as usual, the Um liked to come into the cottage for a snack. Then she would join us in the living-room. After making a tour of inspection and pausing for a few seconds to quiz each person present, she would fold up and settle on the mat with a little thud.

She was a fairly large animal now, but she occupied a surprisingly small space when seated. If she stretched her neck out along the floor she took up more room of course. Paul fetched out his 10-ft rule one evening and found she measured just 7 ft from nose to tail tip but was only 18 in wide. Sometimes she went soundly to sleep in this position. She never slept with the side of her face to the ground like a pony. I supposed that this was because the large eyes in their prominent sockets were vulnerable. Sometimes she seemed to be dreaming. The long ears would jerk, her nostrils would flare and the mobile lips would begin to twitch. The speed of her breathing increased and she would make puffing sounds and blow out her cheeks. Once or twice she made a strange, grating sound which we had never heard her utter when awake. It sounded something like a hastily pushed chair scraping along the floor, an unpleasant and rather desperate noise. We wondered what dreams could possibly accompany this weird croak. When asleep, she never completely closed her eyes. The thick lashes covered most of the eye-ball but there was always a lustrous, unseeing fragment unveiled.

Before we went to bed, Paul would turn the llama out if the night were fine or lead her to the stable if it were wet. Just before she went, I would kneel beside her for a final talk. Usually she did not like having her head touched by anyone but, during these evening chats, she could be most affectionate. She would rub her face up and down hard against my jersey and murmur a whispered *mmm*. She would allow me to rest my cheek against the soft fur of her head and inhale the delicious scent of clean, warm llama. Sometimes, by rubbing round the base of her ears, I could find a ticklish spot which sent her into ecstasies. She would flop her head into my lap and squirm with pleasure, ears flicking and lips twitching. Sometimes she seemed to pass into a deep trance and be quite unconscious of her surroundings, puffing and blowing violently.

'I should think llamas are easily hypnotized,' said Paul one evening as he observed the Um in one of her trances.

It certainly seemed most extraordinary to me to be kneeling on the floor with a writhing llama's head in my lap, while its unseeing eyes flickered and it puffed out great breaths. When the Um came out of a trance, she seemed a little bemused, sat very upright with a slightly offended air and looked round her as though to get her bearings.

'What a very strange animal!' Paul and I agreed.

It always surprised me that such a normally alert animal should sleep so deeply. The keen eyes and huge, efficient ears suggested that a llama needed to be able to detect its enemies at a distance. I wished I knew more about the herd behaviour of llamas in Peru. I often wondered what behaviour patterns were specific to Ñusta and what were common to all llamas.

Her deep sleep gave me the chance to take the kind of liberties which she would not permit when awake. I could get down on my hands and knees and, very gently, push up the twitching lips to examine her front teeth closely. She slept on, unaware of my investigations. I found the teeth most interesting. When she came to us at less than six months old, she had four incisors in the lower jaw. They were long, strong and discoloured. A few months later, I noticed that she had raised another pair of teeth, small triangular corner ones. It was impossible to look in further and inspect the state of her molars, but they sounded highly efficient as she crunched up her pieces of swede and carrot. After she was a year old, we expected her to lose the two middle incisors, as the sheep did, ready for them to be replaced by two permanent teeth. Nothing happened except that the teeth grew longer and stronger, and inclined forward like the teeth of an old horse. It was not until December 1976 that I noticed a change. She was then nearly two and a half years old. I was having a slight investigation one evening while she slept and I saw that there was a small gap on each side of the middle pair of teeth. On closer inspection, I saw that the edges of two new teeth were appearing slantwise through the gums between the first pair of incisors and the second. This had caused a separation of the teeth. As the weeks passed, we watched the slow progress of these new teeth with interest. They overlapped the second pair of incisors in front and were tucked behind the middle pair which, so far, did not seem to be loose.

The Um spent a part of each day in what we could only describe as teeth-cleaning. She would run the outer surface of her teeth backward and forward along any hard, smooth surface. It sounded odd to hear her doing this on the iron trivet in front of the fire. Sometimes she used the edge of her corrugated cardboard toy-box or, if she were outside, a wire fence. When her new teeth were growing, she began to do it on the window pane. We were not pleased about this; she left an ineradicable scribble of finely chased lines on the glass. There was no way to stop her, short of fencing the cottage or letting her indoors. We usually let her indoors, but we were resigned to having scratched windows. From inside it looked most odd to see a llama with its lips squashed against the glass, cleaning its teeth. Also we thought the scribble of lines made an interesting and permanent llama memorial.

As summer progressed, the sky continued to remain cloud-less and the sun shone. The intense heat of June lessened a little in July but it was still extremely hot. There were reports of severe drought in some parts of the country but our fields remained green and our spring continued to run.

In July, Ann, Penny and Amanda rode in their first gymkhana. Neither children nor ponies had competed in a real gymkhana before and the most we could hope for on this occasion was to gain some experience. During the past two summers, they had organized small events on the farm and the four children had ridden against each other on young and unpractised ponies with a surprising amount of professionalism, but I was not prepared for their success at the gymkhana. They collected a host of rosettes between them, Amanda and Penny with Firsts and Ann not far behind. They were even more successful at the next gymkhana. Ann won the Mini Show Jumping on her beloved Dolmen, and so began an engrossing interest for children and parents alike. Paul and I thoroughly enjoyed taking the children and ponies to shows and we entered into the excitement of it all with enthusiasm. We were happy that the pony breeding which we had begun so many years ago was now giving so much fun. We had to take care that our pony events did not interrupt the farm work too much as we set to work to gather our hay and silage.

After a dry spring, the crop was a fairly light one and we had an unusually easy hay-making. Day after day the sun blazed down and we had to rush the green grass to the silo before it turned to hay in front of our eyes. Les, Paul's friend from the war in Burma and leader of the Boys' Club, had emigrated to Australia but his successor, Mick Brown, still brought the boys to camp in Clogwyn barn, together with Julian and pupils from Peers School. That year we had a good contingent of willing helpers to cart and trample the silage. One or two of them had the novel experience of being spat upon by a llama during their mountain holiday. As in the previous year, the Um was busy presiding over our activities with enthusiasm and enjoyment. Some of the young city-dwellers couldn't resist trying to stroke her—with unfortunate results for them. The Um made it plain that she could not tolerate their behaviour, and the lads soon learnt.

The campers went home before we had finished hay-making but Miles came for a fortnight and toiled diligently in the blazing sunshine. We were very thankful for his help. The weather was so hot again that any work was an effort. The heat seemed to make my multiple sclerosis worse, I felt ill and was slower about my work.

We had borrowed an old hay-turner, not very efficient on account of its missing parts but still a wonderful help. With this we were able to handle much larger areas of hay than we usually did. Our hay-making was still a labour-intensive business, as it always had been on our small, oddly shaped and steeply sloping fields, but, with the price of feed still going up and up, we were thankful to garner every scrap. The children weren't much help that summer. The heat made them languid and they spent most of their time in the river.

By mid-August the hay-making was finished. We had cut every mowable scrap of grass and the cowshed was stuffed full. There were several useful loads of hay in the Tŷ Mawr barn and it was all of beautiful quality, grey-green and fragrant with never a drop of rain on it. One of the older farmers told us that we were foolish to keep on mowing when the ground was so dry. There would be no aftermath to feed the animals in the autumn, he said. Perhaps not, but we felt that the grass was withering where it stood and would be better cut and stored in the barn while there was still some goodness left in it.

The mountains basked in continuous sunshine. All of us had developed a Mediterranean-style tan. Sweet corn ripened prolifically in the garden and tomatoes and cucumbers were abundant in the greenhouse. But there was a shadow over the brilliant weather; we were beginning to be short of water. North Wales, which had remained so green when the rest of the country was drought-stricken, was now beginning to dry up too. Some of the trees in the woods began prematurely to assume the shades of autumn. The shallow-rooted birch trees were the first to turn and their leaves became golden. Some of the oak trees, growing over rock, became brown. The ground was dusty and cracked. Bogs and winter's mud were long forgotten.

One animal which enjoyed the weather was the Um. She basked in the sunshine, now perhaps more comfortably without her fleece. She seemed to like eating the sere and sunbaked grasses and she loved the dust. If she found a patch of really dusty soil, she would collapse on to it and roll again and again, with her stick-like legs flailing the air. When she rose, her beautiful, peachy fur would be covered with dust and dirt.

We didn't know the significance of a llama's dust-bathing. Perhaps it served the same purpose as a hen's—to keep external parasites under control. Llamas were supposed to be very prone to these, but as far as I could tell, Ñusta still had none. Ponies roll frequently but any old place will do for them. They don't seem to select a specially dusty spot or a heap of ashes or sawdust, as the llama does. I suppose that rolling for a pony stimulates the circulation and is not connected with parasites.

One day we took the Um with us to a birthday tea-party at Clogwyn, where some friends were staying. Before setting out, I gave her a quick brushing so that she would look her best for the party. She galumphed along with us happily, delighted to be going for an outing. On the way we passed through the field where Julian and his pupils had been cutting bracken some weeks before. When they had finished the field, they had gathered the bracken together and burnt it. Now all that was left was a big pile of grey ash. We noted it as we passed. When I looked back to see if the Um were following, I found that she had noticed it also. I saw a great cloud of dust billowing up into the still, afternoon air and long legs flailing wildly in the middle of the cloud. When she finally emerged from the dust storm with a pleased expression on her face, she was an unrecognizably grey and untouchable llama. I wasn't too proud of her at the party.

The Um liked sawdust, too. She would eat wood shavings, particularly the long, curly ones left by Paul's plane, and would soon tidy up the workshop for him. Sawdust was for rolling on.

After Paul had done a stint of log-cutting with the chainsaw, there was often quite a large pile of sawdust for her. She would soon find it and roll in the middle. Though she was such a clean and fastidious animal in most ways, she didn't seem to mind covering herself in rubbish. The sawdust clung to her fine, dry fibre and was difficult to brush out. We thought she might have been frightened of the chainsaw but she was not. She respected the car when the engine was running, but the ear-splitting din of the saw seemed to attract her. Perhaps she liked the fountain of sawdust which sprayed forth. Whatever the reason, she always behaved in a very excited and stimulated manner when chain-sawing was in progress. As soon as the motor started, if she were anywhere about, she would hurry to the scene and do her best to get in on the act. You could almost hear her saying: 'Oh, what fun! Look at all that sawdust! Why does it fly in the air? Let me get at it!' Indeed, Paul had to be very careful of what he was doing for fear of an accident. When the job was done, the llama would have her roll in the sawdust and then, perhaps, sit down on it to chew her cud.

We didn't do much log-cutting that summer; gathering in our winter fodder was the most important job. Ñusta's second birthday fell just before we had finished the hay-making. With the sun shining and loads still to be carried, I wasn't prepared to devote much time to organizing a llama's birthday party. However, Ann and John would have been disappointed if the day had passed without some cele-bration. Indeed, as on her previous birthday, we felt we had something to celebrate. Now we could look at our large and maturing llama, a well-adjusted and acclimatized inhabitant of the farm, and remember the frail stranger that had come to live with us more than eighteen months earlier.

I baked a quick cake with two candles and constructed a hasty garland of hydrangeas, somewhat ill-made and lopsided compared with the garland of the previous year. Grandma again gave a much-appreciated bunch of sweet peas, John a packet of salted peanuts, Beenie chocolate biscuits and Ann a tin of Coca-Cola. This year we knew exactly what llamas liked. The Coca-Cola was a newly discovered preference. Ñusta was uninterested in most drinks except water (and, of course, cherry brandy) but she loved Coca-Cola. She would sup it from the tin or from a glass with great enjoyment. We didn't know why she liked it so much when she usually refused sweet fruit drinks. Perhaps it was because coca, from which cocaine is produced, is a South American plant.

We worked hard in the hay in the morning, and the birthday tea at Tŷ Mawr was to be a quick affair. Mum walked down to join us and Helen and Brenda came up from Nantmor. We had some friends camping on the farm so, of course, they were invited. Then some more friends called and, by the end, there were about fourteen people at the party. What was intended to be just a token celebration soon turned into quite an event, even though it was rather disorganized. And we managed to carry a few more loads of hay after everyone had gone.

By now our water supply was becoming very meagre. Carneddi and Tŷ Mawr are supplied from a spring on the hillside a hundred yards above Carneddi house. When we bought the farm, we were told that this spring had never been known to run dry. Even in a moderate drought, it continued to deliver a crystal trickle down the bank and into a tub, from which in those early days we carried our household supplies in buckets, and later into a 1,000-gallon storage tank which was constructed just below it. The steady trickle always surprised us because the catchment above seemed to be quite small. Wisely, my father had made certain that the ancient construction of the spring was not altered. We had heard stories of people who had tried to improve their wells and springs, only to find that their water disappeared in the process. We didn't interfere with ours, but just introduced an overflow pipe which led into the storage tank. The spring remained as it had been made long ago, just a niche in the hillside with a small pool of deep, dark water below a covering slab of slate. It was now so overgrown with ferns and moss that it was impossible to see details of the original construction.

Our splendid spring did, in fact, run dry in the long, hot summer of 1959 and we were without household water for about two weeks, except what Paul could cart from the village with the tractor. Now we were afraid it was going to run dry again.

Our farms have every advantage of situation, except for the lack of quantities of water. They have sunshine from early morning until sunset and wide views of mountains, valleys, woods and sea. They are above the level of low-lying mists, frost and floods but, because they lie along the spur, there are no large streams. Most of the time this doesn't matter. Much of the time we have a macs and wellies climate and get a 90-in rainfall in the year. This year we felt as though we would have been glad of quite a large river by the farm.

As the inflow of the storage tank became a thin trickle, we stopped having baths and flushed the lavatory only with water caught from the washing machine. Every drop of spare water went to the garden. Thousands of families throughout the country were doing the same but we knew that when our water supply finally failed, no local authority was going to bring us any more.

At last the inflow of our tank dripped its last drip. We now had about eight hundred gallons of water to last until the next rain came. Paul, ever inventive, was not daunted. There was still a tiny trickle of water flowing down the ditch in Cae Isaf, our lowest field by the main gate. He fixed an old bath in the stream-bed below what, at normal times, was a small waterfall, and led the trickle along a piece of guttering into the bath. This was soon full of sparkling, but possibly polluted, water but any sort of water was a blessing at the moment. We used it to water the ponies on the Tŷ Mawr fields where all the ditches were dry, and each day the children brought a string of nine ponies to drink at the bath. We had already moved the cattle, except the milkers and small calves, to Clogwyn where the stream still flowed.

From the outflow of the bath, Paul rigged an ingenious system of pipes to a 250-gallon oil tank further down the slope. We had bought this tank two or three years earlier, intending to buy our diesel in bulk, but the cost of filling it always seemed more than we could afford and it was as yet unused. The trickle in the stream seemed very small but it went on day and night and, in fact, we found that it was flowing at about fifty gallons an hour. How the water came from that parched earth, we could not imagine. Paul attached a hose to the outlet of the tank and led it into a forty-gallon drum in the tractor's link-box on the road below. Now he could transport quantities of ditch water up the hill to Carneddi and across the fields to Tŷ Mawr. He syphoned some of it directly into the washing machine and I was not reduced to doing the laundry by hand in little dribs and drabs of water as I had feared might be necessary. The colour of the water was not all that might have been desired. It seemed to have picked up some rust and perhaps other foreign matter in its passage through the bath, the tank and the drum, but still it was water. I wondered what it would do to our white shirts, sheets and pillow-slips, but I used plenty of soap powder and the results were not at all bad. Paul kept plastic dustbins filled with ditch water available for use at both Tŷ Mawr and Carneddi, and with this we eked out our diminishing supply of tap water.

'Everyone must wash in hot pond now,' I told the children.

Nobody minded washing in pond water; we were glad to have it.

'Now I know what "dull as ditch water" means,' said Ann, looking at the brownish depths of her washing basin.

I began to boil the tap water for drinking; I felt that the contents of the storage tank must be getting slightly stagnant after so long without refreshment. Now everything indoors seemed a little grimy and I longed to wash and scrub with unlimited hot water. We began to wonder if it would ever rain again.

Then the Nantmor village water supply ran out and the Water Authority set up big tanks. Everyone was carrying buckets. The rushing stream near Clogwyn house dried up and the big pool in front was empty, an occurrence that had never been known before. Friends who were staying at Clogwyn took the opportunity to clean out the pool. With the last foot of mud, they uncovered innumerable eels which they transported down to the river in buckets.

Brenda and Helen were staying at Gwylfa for the summer and, on most days, they would come up to Carneddi for tea. Brenda would bring a jerry can of pure Water Authority water with her. All of a sudden, water had become a precious gift.

Water transport, syphoning to the washing machine and watering the garden and greenhouse occupied a good deal of time. The calves, hens and dogs all had to be supplied with clean water which we carried. The llama's favourite drinking place was the dogs' bowl in the kitchen. She discovered it quite early in her career and drank from it in preference to her bucket in the stable. Contrary to our expectations, she drank frequently though not much at a time. Though such a tidy and particular feeder, she was a messy drinker. When drinking, she would make loud and inefficient slurping sounds and raise her head from the bowl with a stream of drips from her chin. With the dogs and the llama drinking from that bowl, I seemed to be filling it a dozen times a day.

Despite all the extra work that the drought made and the loss of cash that it was causing us, it was impossible not to enjoy those long, sunny days. It was pleasant to wear sandals and minimal clothing from dawn to dusk, to sleep under nothing more than a sheet, to have dry ground for walking and never have to bother about warm jerseys, gum boots or rain clothes. The children took to camping out in the field behind the cottage, Amanda and John in one tent and Ann and Penny in the other, each tent with a dog to guard it at night. I wonder if the children will always remember those cloudless days of their youth, when it was possible to cast aside so many of the trappings of civilization, to sleep on the ground, to run barefoot, to eat out of doors and to rise and go to bed with the sun. Perhaps they will remember being able to ride to the river when they felt like it and return when they wanted in the carefree days of that summer.

The llama found the tents interesting and pottered about the camp when anyone was there, now and again giving an unwelcome twang to the guy-ropes with her teeth. I tried to chase the children into bed at a reasonable hour each night but the evenings were so lovely, still and warm, with the shadows lengthening and the sun sinking behind the hills in a golden glory, that it was hard to get them into their sleeping-bags. Long after they had washed and cleaned their teeth, I found them, in pyjamas and nighties, riding their bicycles over the parched grass or playing football. It was usually the advent of the midges, when the sun had set, which at last drove them into their tents. John slept with his bow and arrow at his side in case of intruders.

At the beginning of September, our friend, Mollie Keen, brought some very welcome visitors to see us. There were Dr and Mrs Turner from Bath and their three children, Dr Ian Herbert from the Department of Zoology at the University College of North Wales and his two children, and Señor and Señora Marcello Hervé from Santiago, Chile, and their two children. Señor Hervé was a veterinary surgeon and, unlike our excellent local practitioners, he was one to whom llamas were just a normal part of farm livestock. He was the very person I wanted to meet.

With the exception of Mollie, none of the visitors had been to Carneddi before, but they were interested to see the farm and especially to meet the llama. They arrived, thoughtfully bringing containers of drinking water with them so that tea and orange squash for so many would not strain our resources. In fact we should gladly have spared our last drop to refresh such distinguished visitors.

I enjoyed talking to them enormously and the time went far too fast. The Um was delighted to have such informed and sympathetic people to interview. She was in a sunny and cheerful mood and showed herself to advantage. Señor Hervé was most approving. He said she was a very fine animal and liked her colour and the fact that we had sheared her. He pointed out with pleasure the strange action of her hind legs, the action that had rather worried us as possibly being abnormal or a sign of weakness. He walked a few wiggling steps over the grass to demonstrate that this was indeed how a llama should walk. We learnt for the first time that the correct pronunciation of the name of our animal was 'yama'. We decided that we would use it when in the right company but that most folk wouldn't know what we were talking about if we constantly referred to yamas.

After tea at Carneddi, we all walked down to Tŷ Mawr to see the old historic barn. Then we went into the cottage. I wanted our visitors to see how civilized and domesticated the llama was. She followed us in.

'If everyone sits down, the llama will sit too,' I said.

Everyone sat down and waited silently and expectantly. I waited too. It was all very well saying confidently that now the llama would sit, but would she? She toured the room, checking on everyone present. Perhaps now she would trundle off into the kitchen in search of something to eat, or perhaps she would decide to go out again. But no, today her party manners were impeccable. She was positioning herself in the centre of her favourite rug and sagging a little at the knees. Then she went down in the characteristic three stages of folding up which is so impressive to watch, particularly at close quarters. There she sat, like a huge swan, composed, confident and evidently realizing that she was the focal point of the party.

Conversation broke out again. Everyone was impressed and charmed by the llama's behaviour. Señor Hervé said that he was quite astonished. He had never seen a llama like that before. In South America, he said, the animals were generally considered to be aggressive and difficult. They did not behave like Ñusta.

'Perhaps that's because of the way they are treated,' said Paul, and Señor Hervé agreed that this might be the case. He advised us to dose Ñusta regularly for fluke and worms and to be on the look-out for ringworm, from which llamas frequently suffered. He did not know if the llamoids suffered from tick-born fever as this was not indigenous to South America. He thought that, having safely reached the age of two, she was likely to be immune to it by now.

There seemed to be hundreds of questions that I wanted to ask, and so many interesting people to talk to that the time passed far too quickly. I was grateful to Mollie for finding so many experts for me. All too soon it was time for the various children to be taken home and we bade the visitors good-bye. It had been a fascinating afternoon and I hoped they would come again.

# 20 The Circus and Some Science

A few days after the visit from the Chilean vet and the other guests, we saw posters in Portmadoc advertising the circus which usually came there each year. One of the listed attractions was 'LLAMAS!'

'I should like to see the llamas,' I said.

'So should I,' said Paul.

Ann and John wanted to see them too. In the end we all went, Beenie, Brenda and Helen, and Amanda and Penny included. We went about an hour before the afternoon performance was due to begin, hoping we should be allowed to see the llamas. The day was beautiful and most of the party wanted to go to the beach after they had seen the animals. I definitely wanted to see the performing llamas perform. I knew something about one particular llama but I wondered if others were like it. I wanted to compare their behaviour and potentialities. I also thought that, if we were allowed to look at the llamas, some of us at least should pay to see the show.

I had not been to a live circus since I was a small child. In those pre-war days, our mother would take Mary and me to see Bertram Mills's Circus when it came to the Nottingham Forest each year. In those days I was enchanted by the magic of it all. Now I was older and maybe wiser but

I wanted to see the performing llamas. Ann, Amanda and Penny said that they would like to go to the show, too, and so it was arranged.

Perhaps one has reservations about performing animals but each case should be judged on its merits. There is no doubt that some dogs love performing tricks and showing off, with all the attendant fuss and praise. A well-trained dog always seems more confident, secure and happy than one which lives according to its own impulses. I think most domestic animals quickly learn and enjoy a routine and if the routine includes a few tricks, the animal probably enjoys that, too. It is when an animal looks stressed during a performance or is asked to act against its nature that the doubts begin to creep in.

We found the circus on the Traeth past the abattoir and the mill, where Portmadoc town gives way to the flat sandy fields that were once reclaimed from the sea. There were caravan trailers surrounding the Big Top and, here and there, a few ponies and horses tethered on the parched grass. Paul parked the Land Rover by the gate.

'Stay here, everyone,' he said. 'Mummy and I will go and ask if we can have a look round.'

The site seemed deserted but at last we found an elderly man who looked as though he might be a clown in mufti. Paul explained our presence and asked if we might look at the llamas. Yes, said the old man, we could. He added that the circus had only one llama, a three-year-old male, and it was tethered to the lorry over there. He jerked his head in the direction. Now we could see the animal, a heap of wool lying on the dusty ground. The man seemed busy and not very communicative so we thanked him and left him to his work. Paul told the children that they could get out now and have a look at the animals.

'Take care not to disturb the ponies,' I told them, 'and don't get kicked. You don't know them yet.'

We all went to see the llama. He was wearing a leather head-collar and was tethered to the lorry wheel by a long rope. He rose to his feet as we approached. He was about the same size as Ñusta but more heavily built, and his fleece was enormous, hanging about him in great, shaggy masses. He did not look as though he had ever been sheared. His fibre was grey and his face and legs dark brown. His neck, top-knot and upper legs were very woolly and, to my inexperienced eye, he seemed to favour an alpaca-style of llama and not the clean-legged and bare-headed type like Ñusta. His ears were smaller, too. He hadn't her type of huge semaphore ears, but he was charming and llama-ish, though without Ñusta's elegance.

We approached him slowly and he came over to us to the limit of his rope. I had a few crumbs of dog biscuit in my pocket so I offered them to him. After a little hesitation, he scrunched them up. Ann gave him a Polo and he enjoyed that, too. We watched for warning signs of spitting but the animal seemed very amiable. I was interested to notice that his toenails were very overgrown. This, I supposed, was how Ñusta's would have grown if we hadn't trimmed them. The circus llama's nails were so long that they curved into a circle flat on the ground. They did not seem to cause him discomfort at the moment but I felt that they could give trouble later.

The girls now went to see the ponies and two or three goats which were tethered along the fence. I turned to see where Brenda and Helen were. Then, thump, something heavy struck me and I was on my knees on the ground. For a moment I was winded. The llama had come up while I was looking round and had played Ñusta's jumping game on me. His knee must have struck me in the midriff and I was quite taken by surprise. Paul hurried up and the llama moved away. In a moment I was all right again. I stood up but I felt a fool. I was supposed to be the llama expert and I had been caught unawares. I looked hurriedly round to see if any of the circus staff had witnessed the incident, but none had. There was no one in sight. I was thankful because I didn't want any fuss. The incident had been my own fault; the llama was not being aggressive. It had just done what llamas sometimes do, and it had made a fool of me. I dusted myself down and tried to look like a mature animal-lover, to whom nothing had happened. I gave the llama another Polo to show that there was no ill-feeling.

Next we had a walk round the tethered ponies—two or three Shetlands, a big white Liberty horse, a spotted horse and one or two others. They were all in excellent condition and, unlike the llama, had neatly trimmed hooves. Then Paul, Brenda and Beenie said they would go to the beach with the younger children while the sun still shone. The girls and I waited expectantly for the performance to begin and we were in our seats early.

After the show, I felt glad I had taken the children to it. The circus was a very small one compared with the Bertram Mills's of my youth but it was very professional circus. There was knife-throwing, rope-spinning, fire-eating and a good fast-moving clown act. The fire-eater lay on broken glass and a bed of nails while a large lady stood on his chest. There were no elephants and no big cats, and these I did not miss. It was during such acts that my doubts about the ethics of seeing performing animals are aroused. The pony and dog acts were clever and the participants seemed to be enjoying themselves. Our girls loved the broad-backed Liberty horse and, perhaps, were dreaming that it was they who were riding it. I was on the look-out for the llama.

When his turn came, he was led into the ring on his halter. Small jumps had been set up round the ring and now he galumphed round enthusiastically, leaping each in turn while his handler stood in the middle holding the end of the halter rope. His name, we learnt, was Larry. It looked most odd to see this huge mass of hair undulating over the jumps. I was surprised how well he jumped. Ñusta had always indicated that she found jumping impossible except in extreme circumstances. We had never tried to change her mind. It suited us better to be able to confine her easily though Ann had tried to teach her to jump. Ñusta, however, always stopped to investigate even an ankle-high obstacle. Then she would step over it, one leg at a time, if the reward on the other side were sufficiently tempting. If the jump was more than knee-high she refused to attempt it. Now we saw Larry clearing jumps as though he enjoyed them. Evidently llamas could jump if they wanted.

At the end of the act, Larry knelt before the audience and then lay down. To make him kneel, his handler touched him on the front legs with a cane. I found this interesting because I had already discovered that Ñusta would begin to kneel if I touched her forelegs, though I had never tried to train her. Now Larry, having lain, was unwilling to get up again. He had galumphed merrily round the ring, jumped his jumps and was now ready for a quiet sit. The lights and the audience did not disturb him in the least. His behaviour reminded me so much of Ñusta's that I couldn't help smiling. His ears went back as the handler tried to rouse him. We wondered if he were going to spit but he didn't. Finally he rose to his feet with an 'oh, all right' expression on his face and followed her out of the ring. We all applauded enthusiastically.

At the end of the show, children were allowed to ride the circus horses round the ring for 10p a time. Ann, Amanda and Penny were soon in the queue for rides. Our children, who could ride all day for nothing, were soon mounted and progressing round the ring at a stately pace, each led by a clown and each with a big beam on her face. One might have thought that it was their one and only chance of sitting on a horse.

We had all enjoyed ourselves immensely. I was struck by how hard the artistes had worked for what must have been very little recompense. The Big Top was a small one but it was less than half full at that first performance. The performers had all done several jobs. The spangled lady of the trapeze had put on her coat and served candy floss in the interval. She had thrown knives, spun a rope and had her cigarette cut in half by the whip. Others had announced, moved props, regulated the music and taken part in several acts. This small circus was cosy and intimate and there was no doubt that the artistes possessed quite exceptional skills. I was glad that the children had seen some real, traditional circus at close quarters.

When we met the rest of the family, our enthusiasm made them regret that they hadn't come with us. They wanted to see a jumping Um. John would be fascinated by the knife-throwing and the rope-spinning, and so it was decided that they should go on the following day. The next afternoon Paul took John, Beenie and Helen to the circus and, of course, the three girls wanted to go again, so they did. I had seen what I wanted to see so I stayed at home to get on with my work.

Later in the evening, the family returned. They had all enjoyed the show. Little Helen was so excited by the final pony-riding that she had rushed into Gwylfa shouting: 'Mummy, I've ridden on a clown with a pony leading it!' John was wondering how he could acquire a set of throwing-knives. Ann had had another ride on the Liberty horse. Beenie and Paul were impressed by Larry's hurdling.

Paul had also talked with the talented spangled lady. She was Miss Jessie Fossett, circus born and bred. She told Paul that Larry belonged to her and that she had had him for a year but had not had much previous experience with llamas. She was interested to hear that we had a llama too. They discussed the question of breeding from Ñusta and she gave Paul the address of circus folk who might have a stud male. Larry could not be used for breeding because this might unsettle him for his act. She asked Paul's advice about shearing him but Paul said he thought that it was now too late in the season. They discussed the overgrown toenails and Paul said that we would be glad to cut them, if we were in Portmadoc again before the circus moved on. He seemed to have thoroughly enjoyed the skills and atmosphere of that circus.

After breakfast next morning, when Ann and John had gone to school, Paul said: 'Shall we go and cut that llama's nails? The circus will be gone tomorrow.'

'I'm game,' I said, 'but how will we restrain it? I should think it's stronger than our Um and perhaps hasn't been handled so much. She doesn't like toe-trimming and Larry's going to hate it.'

We got out our old copy of Miller and Robertson's *Practical Animal Husbandry* and looked up the chapter on the restraint of animals. Quite a variety of methods of casting livestock were illustrated but none of them seemed to suit our purpose. Some of them needed special hobbles and other equipment that we didn't possess, all of them seemed complicated and we hadn't used them before. We weren't going to be able to cast the llama with a book in one hand and a rope in the other.

'I think we shall just have to get the circus strong men to help us overpower him,' I said, 'then we can tie him up like a sheep. I'm sure we can never do him standing like the Um. He'll get all worked up and kick, not to mention the spitting.'

Paul thought that was probably right.

We collected the sheep hoof-clippers, a sharp knife in case of need, a length of rope and put on our oldest anoraks to protect us against possible spitting.

When we arrived at the circus ground, Miss Fossett came out of her caravan. She was pleased that we had come to trim Larry. The other members of the circus gathered round and we discussed the best method of tackling the job. Larry stood by thoughtfully, unaware of what was about to happen to him. It was agreed that the simplest way was to tie Larry's feet. The rope-spinner went to get his spinning-rope and then the four men—Paul, the rope-spinner, the fire-eater and the younger clown—seized the llama and collapsed him on to the ground before he knew what was happening. In a moment, two of them had tied his four feet firmly together and he lay on his side, looking up with a startled expression on his face. He did not spit. I was ready with the clippers. It was none too easy to get at his feet; they were all bunched together by the rope and covered by big locks of his wool, but I managed to sort them out and began to clip. The horn of the toe-nails was extremely hard on account of the weeks of dry weather and it took all my strength to close the clippers. Paul and the circus men held Larry down to prevent his struggling. With each cut, I positioned the clippers carefully to try to avoid the quick in each claw. The situation of the quick was rather academic as far as I was concerned; I did not know the anatomy of the foot. However, I tried to cut so that the feet would look right when the animal was standing. I trimmed the ponies' and sheep's feet on the same principle and generally found that if a thing looked right, it was right. Now, perhaps, I erred on the cautious side but I couldn't risk any injury to Larry. Great hoops of overgrown horn came off, even with my timid cutting, and now he would look better and be more comfortable.

The job was soon done. The rope was released. Larry struggled to his feet and stood looking rather ruffled. One of the circus men fetched some brown bread crusts to cheer him up after the ordeal. I looked with pleasure at the neater feet. I had wondered how we would manage, but the job had gone well and was now done. Miss Fossett thanked us and asked what she owed but we said it was nothing; we were pleased to be able to help. We parted on friendly terms. As we drove home we were smiling happily. We had enjoyed our brief contact with the circus and had gained another interesting piece of llama experience.

When we reached home, we wondered how the Um would react to the smell of Larry on Paul's coat. She greeted us as usual. We encouraged her to examine the coat and our hands but she was quite indifferent to them. There was no secret message for her there. All that interested her was the possibility of concealed chocolate in our pockets. This raised a few doubts in our minds. How would she react to a male llama when we tried to breed from her? After being separated for so long from others of her kind, did she now regard herself as human? Would the story of Chi-Chi and An-An, those pandas from England and Russia who wouldn't breed when brought together, be repeated? We knew that the breeding of llamas was a complicated procedure and that the results were uncertain.

By now I had read, and done my best to absorb, the contents of those interesting papers which I had received from the library of the Royal College of Veterinary Surgeons during the past spring and summer. Much of the information was so specialized that it was beyond my comprehension, though my cousin had helped to clarify some of it.

What I really needed was an *Idiot's Guide to the Physiology of the Llama* but none was available. Here and there I could understand the work of the researchers and some fascinating facts emerged.

The papers on breeding were the easiest to understand and these were the ones which interested us most. It seemed that there had been a good deal of research into the breeding of the South American Camelidae, particularly alpacas, carried out in Peru, Bolivia and the United States during the past fifteen or so years. These animals are very important to the inhabitants of the Andean *altiplano* whose economy is based on animal husbandry. Some of the large numbers of llamas and alpacas in the Andes now belong to middle-class owners and associations that have emerged as a result of land reform, but most belong to small owners and *campesino* communities. In the past, the Incas had selection programmes for the improvement of their herds and they practised rotational grazing to ensure food supplies. These good practices of management were lost with the conquest of the Inca Empire and have not been revived. Now researchers hope to improve the standard of llama husbandry and make the mountain communities more prosperous. It has been found that llamas and alpacas have characteristics so different from other ruminants that animal husbandry techniques developed for other species can't be used for them. Where alpacas are managed like sheep, the production is very low. We read that only fifty per cent of the females produce young each year, and one authority put this figure as low as five to twenty per cent.

As a farmer and llama owner, I was enormously interested in the information contained in the papers. The facts which emerged, shortened and simplified, seemed to me to be as follows. (I use the term 'llama' to cover both llamas and alpacas, the domestic animals of the Andes.)

The mating behaviour of llamas is different from that of most range animals. Cows, sheep and horses have regular oestrus cycles during at least part of the year, and ovulation, the shedding of eggs from the ovary, takes place spontaneously when the animal is ready for mating. The llama, however, is on heat for weeks at a time and ovulates only after it has been mated. Cats, rabbits, ferrets and mink have the same breeding pattern. Female llamas may reject the males by running and spitting but, if the female is receptive, the courtship lasts for only a few minutes. The female lies down for mating, which may take twenty to fifty minutes. The mortality rate of embryos is high during the first month and only fifty per cent of them survive the first thirty days. The reason for this is not known, but it may be connected with poor nutrition. No multiple births have ever been recorded. Males will serve up to twenty-five females, and there is a higher birth-rate when two groups of males are used alternately at seven-day intervals. The fertility of the males is low.

Breeding is mainly confined to the wet season in the Andes. Llamas breed all the year round in zoos though there is a peak period for births in June, July and August in the Northern Hemisphere. This is typical of many domestic animals and suggests the llama's long history of domestication. Guanaco and vicuna, in contrast, have a well-defined breeding season.

Alpacas are sheared annually. They live for about fifteen years but their teeth begin to deteriorate rapidly after the age of ten and then they should be culled.

'We shan't cull you, Um,' I told the llama. 'We'll take you to Mr Chase.'

Mr Chase was our dentist. He had kindly lent me his copy of *Comparative Dental Anatomy* and I had spent some happy hours studying the teeth of the various species in their bewildering variety. It was interesting to see the dental formula of llamas and camels. At the moment, Ñusta had six incisors in the lower jaw and none in the upper. I had imagined that her dentition would follow the pattern of sheep and cows which have eight incisor-like teeth in the lower jaw. These meet against a hard pad above and are quite efficient for grazing. Now I found that llamas and camels, unlike the true ruminants, have two incisors in the upper jaw and a pair of canines in both upper and lower jaws. This was the formula:

$$\text{Incisors } \frac{1}{3} \text{ canines } \frac{1}{1} \text{ premolars } \frac{3}{2} \text{ molars } \frac{3}{3}$$

It was clear that Ñusta's dentition was not yet complete.

Besides having unusual teeth and breeding habits, I found that the llama also has unusual blood. I had asked the library for several papers on the animal's special adaptability to high altitudes. It seems a fascinating subject and no doubt it is, but one which is rather beyond the level of my education and, perhaps, my intelligence. This being so, it may be risky to say anything about it but I am tempted to try. These are some of the facts which I think I have understood.

The blood of llamas coagulates faster than human blood. The red cells are smaller in size and elliptical in shape, which may help in the transport of oxygen, though other mechanisms play an important part in this adaptive phenomenon. The haemoglobin, which gives the blood its red colour, has a higher affinity for oxygen, and the tissues of the lungs are better adapted to extract the oxygen from the air. At high altitudes the llama, unlike most other animals, does not produce extra red blood cells from the spleen to compensate for low oxygen in the air. Also, unlike other animals, the rate of its heart-beat is not much increased by low barometric pressure, nor does the animal pant. The llama, it seemed, was truly a very special high-altitude animal.

In my reading I could find no more information about those odd, crusty patches on each side of Ñusta's hind legs, those 'glandular spaces in the metatarsal region'. I was longing to know their purpose. Ñusta rubbed these areas on her thighs when she made those strange motions of twiddling her back legs together like a fly. There might be a clue here but I couldn't quite see what it was and the function of those glands—if glands they were—remained a mystery. Perhaps Señor Hervé could tell me more about them if I saw him again.

We knew we still had a great deal to discover about llamas. Each piece of research seemed to end in another of Professor Allen's question marks. Where would we find a suitable male for Ñusta? We knew now that breeding llamas was no simple matter and might have disappointing results. Would we have problems, or would we be blessed with the beginners' luck which had so often attended our ventures? But we did know that llama-ownership, even after two years, was a state that we enjoyed very much indeed.

# 21 The Third Year

The winter of 1976–7 was a colder one than we had had for several years. There were deep falls of snow and the children were able to have days of sledging and fun on Paul's ancient skis — with a broom handle and a hoe in place of the missing ski sticks. John and Paul ascended the snowy summits with their ice axes and did some fine climbing on Moelwyn Bach and Moel Hebog. It was a good performance for a boy of eight, and not a bad one for a daddy half a century older. The snow did not lie for long and I was thankful for the sake of the sheep. They were able to get a bite of grass before the next fall came. In between the snowfalls the weather was wet. The dazzling heat and drought of summer seemed very far away, though we remembered it as we fed fragrant hay to the cows, hay which had been made under blue skies months before. It was reported to be the wettest November to March on record.

The llama seemed to take well to the snow, and the crisper and drier it was, the better she liked it. She would galumph down the hill after the children's toboggans, showing great enthusiasm for the sport. I liked to watch them, the children, dogs and llama enjoying the snow, and sometimes I took a turn on the sledge myself.

I was now deeply engrossed in writing the llama book. Of all the writings I had done in the past, I think I enjoyed this the most. I found it difficult to tear myself away from it and try to concentrate on something else. Now I was neglecting many of my farm jobs but Paul, Beenie and my mother valiantly tried to fill in the gaps. I employed extra labour to help out whenever I could. My aim was to finish the book in twelve months and not to take the long five years which had been needed to write each of the other two books. Now the weather was so wet, raw and bleak that the Writing Hut was no longer an attractive retreat. Mother suggested that I use Mary's room. Here I could be warm and leave my writer's paraphernalia spread about without disturbance. It was a room of sad memory, but I knew that my sister would have been glad for the remembrance of severe illness to be dispelled now by an atmosphere of creative work. Here I retreated with enthusiasm whenever I had the chance. In the afternoons I often had to work against the noise of children's television. The rooms at Carneddi were not at all sound-proof and I found this a great distraction. Sometimes I resorted to wax earplugs but they felt strange and made my head ring. Often I would give up and turn my attention to something else, but I always longed to get back to my writing.

It was a great pleasure to write about that charming, fluffy animal. By now I loved her dearly. I found that there was something immensely attractive about her. Whether it was her elegant but unusual appearance, the softness of her wool or her faint, delicious smell, I did not know. Perhaps it was her character that made her specially appealing, her readiness to listen and respond, even her moods, when sometimes she was ready to spit or throw a tantrum while at other times she was courteous and affectionate. Or it may have been that sort of llama sense of humour which made her stalk up on silent feet to give someone the fright of their lives.

I know that any suggestion of anthropomorphism, the attributing of human characteristics to animals, produces howls of disgust from many. Nevertheless, when I looked into Ñusta's eyes, now nearly on a level with my own, I had the uncanny feeling that there was someone in there, trying to communicate, a someone, perhaps, who was very simple, naïve and trusting but yet one who could put two and two together in a modest kind of way, who experienced a wide range of emotions and who lived her life with enormous zest. The llama's range of toots and *mmm* sounds was great. She was almost always mute when strangers were present but, with the family, she used her voice to express her meaning, and what she meant seemed clear to us. There was something very companionable about being able to share a swig of coke or a bit of chocolate with a llama whose face was on a level with yours and who had accepted it politely and with enjoyment. I couldn't help feeling that there was something a little more than animal about her.

We derived a good deal of amusement from the process of finding out what llamas like to eat. After the first desperate days when she would eat nothing, the range of her tastes had extended. The first real surprise had come when she showed a liking for Easter eggs. After that the variety of her choices grew and grew. This reminded me a little of finding out what Tiggers like, except that the process worked in reverse. From liking nothing, the llama developed a taste for many unusual foods, though Roo's Strengthening Medicine was never among them. Most of the discoveries were made by accident and quick action often had to be taken to protect expensive human food. We found that llamas like bread but not bread and butter. They love biscuits but dislike cake. They like any breakfast cereal until the milk is on it. They adore coffee beans, and tea straight out of the packet but reject the drink made from either. They like a sweet liqueur, Coca-Cola and occasionally Ribena, but dislike wine, beer and most fruit drinks. If Ñusta should accidently taste something she didn't like or one of her favourite foods which was in any way tainted, she would open her mouth and hold it open for several minutes in a gesture of the utmost disgust. This mannerism looked very strange. The corners of her lips were drawn back slightly and her jaws now had something of a beak-like look.

Ñusta liked all sorts of fruit, also tomatoes, but she preferred the skins of bananas and oranges to their insides. Much as she loved apples, she would never eat a core which had been bitten by someone else. She was choosy about chocolate or a biscuit; you had to break a piece for her or she would have none of it if the other half had been bitten.

She was fortunate to have good supplies of fruit. Our greengrocers in Portmadoc, the Prichard brothers, were extremely kind in giving large bags of bruised apples or other damaged fruit and vegetables for the llama's consumption.

'It's Llama Charity Week,' David or Bryn would say as they handed over an extra large bag of goodies.

They thought I was joking when first I asked them if they could supply some llama food, but they still obliged. Those bags of apples and boxes of carrots were invaluable in coaxing Ñusta to eat during her first weeks with us. Later they were a weekly treat for her. She had many a good bowlful of delicious fruit and veg after I had washed and trimmed her bounty to the standard which she always required. The Prichard brothers became resigned to the fact that they numbered an impecunious llama among their customers.

In the late summer, we found that Ñusta loved the small, black plums which grew in profusion on the trees by our cottage. She would help herself to those that she could reach on the lower branches and then munch up the fallings. I wondered if so many plums would upset her digestion. I also wondered what would happen to the stones. We soon found the answer to that question. The llama came in to sit with us one evening after an afternoon's plumming. She seated herself on the mat and meditated quietly for half an hour or so in her usual way. Then she decided to chew her cud. Up came the bolus and left, right, left, right went her jaw. Then there was a slight scrunching and, pop, out came a plum stone on to the floor. The stone was clean and bleached-looking without a vestige of pulp left on it. It had been subject to the powerful digestive processes of her rumen. By the end of the evening there was quite a litter of clean little plum stones on the floor in front of her.

A llama had made quite a difference to our lives. There were obvious signs of her presence in the cottage, signs which unfortunately added to its shabbiness but which made me smile when I saw them. Perhaps, I thought, they were unique and even faintly prestigious. The window panes had a scribble of lines incised by llama teeth. The cotton folk-weave table-centre was frayed at the corners where she had chewed it. The handsome rag rug, which I had made in the evenings while I was sitting with my mother soon after my father's death, looked somewhat moth-eaten. The rug was Ñusta's favourite resting place. It was warm and thick and she liked its woolly texture. As she sat, she would sometimes idly tweak out a small piece of woolly rag and eat it. The rug was easy to repair because it was home-made but she was making more work for me. I would put the toy-box in front of her to divert her attention.

The llama's presence had made a difference to other homes too. After a night of gales, John brought a chaffinch's nest to me. He had found it on the flagstones outside the cottage, and we thought it must have been dislodged from the clematis which grew over the doorway. The nest was the most beautiful I have ever seen. It was made of dark green moss and a little sheep's wool with flakes of light greyish-green lichen round the sides, giving it almost the appearance of a jewel-studded crown. Three or four gingery-coloured feathers, dropped from the breasts of our hens, were woven into the upper edge of the nest in an upright position, their natural curve forming part of a little feathery roof over it. The lining contained a few black hairs, perhaps from the cattle or the ponies but the main filling, deep, soft and luxurious, was beige llama fur. This chaffinch, I thought, must be an expert in design and a devoted mother to have provided its eggs with the most expensive kind of wall-to-wall carpeting. This seemed the last word in modern furnishing. Now the nest made an exquisite ornament on the mantelshelf at Carneddi; it was too lovely to discard.

The spring of 1977 was slow in coming. In May we could hardly believe that summer was nearly here. The weather was still cold with frosts at night or heavy downpours and strong winds. We were well behind with our farm work but I had nearly finished writing the book and, with that task done, I should be able to set to work to catch up again with renewed vigour. By the end of the month the book was nearly complete and then there was the shearing to do and silage to make. We decided against shearing the Um that summer. Her fleece had grown considerably during the past winter but it was still not quite so long and luxuriant as it had been before we sheared her. I think we all wanted to see her fully garbed in her trailing draperies again.

It looked as though there wouldn't be enough grass to make hay that year. We had kept the sheep on the lower fields later than usual because of the long cold spring. After we put them up on the mountain a group of hardened jumpers, our own and our neighbour's sheep, had persistently come down to graze on what should have been our hay fields. No matter how often we put them up, they were back again the next day. We also had more cattle than usual that year and some of the athletic heifers would burst their way through to eat the better grass. We went round building up walls and plugging gaps but in the meantime the best of the grass was gone. There was still enough left to make silage and we set to work to fill the silo. There was an occasional day off to take the children to a gymkhana or One Day Event, outings which Paul and I thoroughly enjoyed. It was most exciting to see them acquitting themselves so well on our little homebred ponies.

But the summer was racing by.

'We really shall have to do something if we're going to get the Um mated this year,' I said to Paul. 'It would be so lovely to have an Umleite.'

It was now mid July. Normally we took our mares to the stallion in late May or early June so that when they foaled eleven months later, the spring grass was growing nicely.

Llamas also have an eleven months' gestation but we had read that most llama births in the Northern Hemisphere occur during June, July and August. We thought it possible that a mating might be more successful if we timed it so that Ñusta might give birth during that period in the following year. At one time we had intended to take a few days' holiday and make a tour of all possible males. Now it seemed that that was out of the question. We were so behind with our farm work and there was no one who could help Beenie to manage in our absence. Also the price of petrol and diesel had risen so much that we no longer took to the road without a very definite purpose. We had seen the male at Colwyn Bay Mountain Zoo but felt he was too small to be ideal. He was a plain white in colour and we should have preferred to use a parti-coloured animal.

'Some male is better than no male at this stage,' said Paul, 'and he's conveniently close.'

I quite agreed, but still I wanted to have a more impressive animal as a mate for Ñusta if it were possible.

'Let me write to Chester,' I said. 'When I went to the zoo on the school trip a couple of years ago there was a handsome male there. I only saw it in the distance but it looked quite big and had a black head. Perhaps a large zoo won't want to be bothered with us but it's worth a try.'

So I wrote to the Director-Secretary again and received a very courteous letter in reply. He would be pleased to help us and we could bring our llama to the stud male at the zoo. He drew our attention to the fact that the mating of animals could be a tricky business and said that the proceedings must be at our own risk. By this time Paul and I were fairly experienced at taking risks. We didn't know quite what to expect in this case but we did know that a bit of common sense and forethought could avert most disasters. The fee would be £20, including keep if we wanted to leave the llama at the zoo. Paul and I were delighted. We considered the fee very reasonable for such an exotic animal. Pony stallion fees ranged from twenty to thirty-five guineas and were sometimes much more. A telephone call and all the business was fixed; we weren't going to let the grass grow under our feet any longer.

# 22 To Chester Zoo

We set out for Chester early one overcast morning at the beginning of August. We had put several loads of grass on the silo the day before and hoped it would not overheat in twenty-four hours. We wondered if Ñusta would travel as well now as she had done two and a half years earlier during her journey from Harrogate. I had been meaning to lure her into the trailer a few times and feed her there so that she could grow accustomed to it, but there never seemed to be an opportunity. Now we put on her collar and lead and took her across to the trailer. She was unwilling to set foot on the ramp but there was no time to waste. With an animal the size of a llama, this did not present much of a problem. Ann went into the trailer first with a pan of nuts and the lead. Paul and I clasped our hands behind the llama's haunches and in a second she was inside. We closed the ramp and let Ann out through the groom's door.

Amanda and Penny were coming too and all the children climbed into the back of the Land Rover. We were off. After a mile or so we stopped to see how the Um was faring. As I had hoped, she was sitting in the swan position and seemed quite composed. When travelling with an animal in the trailer I usually felt a little tense, imagining how every bump in the road, sharp corner or noisy lorry must be affecting it.

I was thankful that our precious llama seemed to travel so well.

We were in good time and hoped to be at the zoo before noon. Though it was still quite early the children had begun to delve into the picnic basket. I was interested to look out of the window and note the progress of hay-making on the high farms of the Denbigh moors. Some fields were a bright lime-green where the hay had been cleared; in others damp swaths were lying. The clouds were low now and rain looked as though it might come later. Then, up the last hill before we reached the Forest, the Land Rover began to labour. Paul changed into a lower gear. I could see wisps of steam whirling away from the bonnet.

'Oh drat it! The radiator's boiling!'

Something was wrong. Paul had filled up with water before we set out. The Land Rover was at its tricks again. Second-hand when we bought it and after fourteen years in our service it was now distinctly unreliable. Most of the time it went perfectly but on occasions of importance it was apt to crack up. On the brow of the hill we stopped to let the engine cool and consider the next move. The children began on their picnic in earnest. I went to have a look at the Um who was now on her feet and looking out inquiringly over the ramp.

'Slight technical hitch,' I told her.

When the engine had cooled a little we coasted down the hill to the next tiny village. Now there was an overpowering smell of diesel. Paul drew up outside the one and only garage. He lifted the bonnet while the engine was still running and we saw precious diesel oil spewing out on to the road. The fuel pipe had fractured.

'It would happen today!'

In a way I wasn't specially concerned. This had all happened before. Paul, although no lover of engines, was an effective mechanic in an emergency and I knew he would patch things up somehow. We would get to Chester in the end. We had never before been stranded on the Denbigh moors with a llama on the way to its nuptials but there was an element of the ridiculous in the journey and a breakdown now seemed in keeping with the whole hilarious business. We all baled out and Paul enlisted the help of the garage proprietor. He was astonished to see a white face with long ears craning out of the trailer and charmed to watch it picnicking off a few of our biscuits. The garage was only recently acquired and not yet equipped for repairs but the proprietor set to work to mend the fuel pipe as best he could. He had no radiator sealing compound so we filled up with water and hoped we would manage to get to the next garage. All this took time. The children and I went for a walk up the lane and gathered a few wild flowers while Paul and the kindly man struggled with the repairs. The Um craned out of the trailer to see us go. I think she would have liked to join us on the walk but we thought it better not to risk more delays.

At last the repair seemed done and we set off again for the next garage in search of sealing compound for the radiator. Before we reached it the fuel pipe fractured again. We arrived at the garage in a haze of diesel oil. While Paul and the mechanic set to work, I went to a nearby grocers to buy fruit and sweets. As I was about to leave the shop, I thought of the Um.

'Have you any of the kind of biscuits that llamas like?' I asked the grocer.

He looked puzzled and surprised.

'What are those?' he asked.

'Something crisp and sweet,' I said. 'There's a llama in the trailer outside and she's rather bored with waiting.'

I selected a packet of Sweet Tea biscuits.

'I must see this llama,' said the grocer.

He followed me outside and the rest of the customers came, too. The llama leaned graciously over the ramp and accepted the Sweet Tea biscuits with relish. Quite a crowd had gathered although it was twelve o'clock on a Wednesday morning. I said my little piece about the fascinating qualities of llamas and the fact that we were taking this one to Chester Zoo. Everyone seemed most interested.

'Would she eat these?' asked an onlooker, showing me a box of peppermint sweets.

'Try her,' I said.

The man offered a sweet and the llama lipped it delicately from his palm. Yes, she did like Tic Tacs. Next I held out the tissue box from the front of the Land Rover to her. She whipped one out and consumed it in seconds. The crowd watched in speechless silence as tissue after tissue, mauve, yellow, pink, blue, mauve, yellow, pink, blue, went the same way.

'I've never seen anything like it,' said a bystander finally.

At last the Land Rover was ready and we could set off once more. We waved farewell to the onlookers who wished us luck with our mission. We were on the road again.

We had no further troubles and arrived at Chester Zoo, late but undoubtedly in good order. We drove to the Time Office as instructed and inquired for the Curator of Mammals, Mr Wait.

'Shall I say: "I Work, you Wait?"?' said Paul in an undertone to me (our unusual Orkney surname being Work).

'Certainly not!'

Mr Wait arrived and guided us round outside the perimeter of the zoo so that we could enter near the llama enclosure and avoid the crowds. We drew up and went to inspect the animals. There were three males; one yearling by itself in a pen, an adult llama in another enclosure and the blackheaded male, which I had seen before, with two females. Mr Wait suggested that we put Ñusta with the lone male. It was a fine animal with a very long fleece and black markings on its head and neck. The feature which I didn't like about it was its eyes. They were a very light blue and gave the animal a strange, staring appearance. It was a proven sire, Mr Wait told us, but I still preferred the black-headed animal. It was agreed that we should try Black Head.

We lowered the ramp and let Ñusta descend. She followed me meekly across the concrete yard and into a spacious service room from which the doors of the llama houses opened. Then Mr Wait opened the door which led into the quarters of Black Head and his wives, a large loose-box littered with clean straw. From this an open door led out to the llama paddock. The paddock, we noticed, was also occupied by a number of wallabies. Outside there was a large porch or 'umport' with a concrete floor and perspex roof. The three llamas were at a little distance from the porch.

Now was the moment of truth. Now we should find out if Ñusta regarded herself as animal or human and if she had forgotten all about llamas.

The three animals in the paddock noticed the presence of strangers. They advanced towards us at a stately pace. The four children, Mr Wait, Paul, the Um and I all stood in the 'umport' out of the wind and watched their approach. Ears forward, Ñusta looked at them intently for a moment. Then her attention was diverted by the sight of a wallaby going *boing, boing, boing* across the middle distance of the paddock. She seemed to find it fascinating. Then the zoo llamas arrived and Ñusta advanced to meet them. The four animals circled one another like the dancers in a slow-motion minuet. The meeting appeared to be extremely polite and formal and Ñusta seemed quite equal to this new social situation, indeed one might have thought that she had been meeting strange llamas all her life. The animals circled slowly, heads high, lower teeth just visible, a slightly supercilious expression on their faces. They ignored the humans standing in their 'umport'.

Black Head, the male, was an inch or two taller than Ñusta but the females were the same height. All the zoo llamas seemed to have longer necks than Ñusta but it may have been that they were less plump and woolly than she was. Black Head's head was really a dark chocolate brown and he had huge lustrous dark eyes. Now he reached out to sniff Ñusta's tail. His manner was polite and tentative. You could almost hear him saying: 'Excuse me.' Ñusta stopped in her tracks and her ears went back. Black Head made a half-hearted jump at her but she walked forward again so that he missed. Ñusta came over to me for protection and the male retired a few paces, looking as though he had been misunderstood. The two females, bored now, went out into the paddock again. Ñusta ventured a little way after them to have a closer look at the wallabies. The male lay down.

'The research papers say that if llamas don't mate within the first four minutes, they are not going to,' said Paul.

Three-quarters of an hour passed. The grey clouds let go a sprinkling of rain and the females came hurrying back to the shelter of the 'umport'. Black Head stood up again. Then we all just stood there and watched the rain.

Had our journey been in vain? Was Ñusta in a state of anoestrus?

'Let's try the other male.'

The research papers had said that the best results were obtained when males did not run continuously with females. Perhaps Black Head was inhibited by the presence of his wives, and the lone male, which Mr Wait had originally suggested, was the more likely animal. At this stage I was quite willing to try the wall-eyed llama.

We led Ñusta away and the zoo llamas did not object to our going. Ñusta came along willingly. There were now two keepers in the service room. They looked at our llama with appreciation and seemed impressed by her good behaviour and beautiful appearance. We led her into Wall-eye's loose-box and the children crowded in, too. Wall-eye was out in his yard and I led Ñusta towards him, released her lead and stepped back to the shelter of the doorway. It was now raining steadily. The male swung round and came galumphing over. He mounted her immediately and Ñusta, taking fright, came rushing to me with the male prancing behind on his back legs. Quick as lightning, Paul stuck out his foot and tripped up Ñusta's forelegs so that she collapsed to the ground in the correct mating position for llamas with the male on top of her. Success.

Unfortunately the pair were rather awkwardly close to the wall but the male seemed to be managing. The children squeezed out of the doorway into the rain for a better view. Now we heard the strange 'singing' of the male llama at mating time, a harsh continuous umming which never stopped throughout the procedure. This lasted for about forty minutes. Wall-eye's fervour did not abate during this time. His neck was stiffly arched, his ears forward and there was a glazed look in his pale eyes. He ummed with such a feverish intensity that he was soon frothing at the mouth. Ñusta, on the other hand, seemed strangely detached. She was down in her favourite tea-cosy position, ears forward and a sweet expression on her pretty face. She looked about her with interest, seemingly unaware of the stormy behaviour of the male on top of her. Once or twice she made as if to rise but Wall-eye's weight was too much for her so she sat on. A bison came to peer over the wall which separated his paddock from the llama's yard. Ñusta turned her head to look at this strange creature. The bison, seeing nothing much of interest, turned away and went about his own business. A little knot of zoo-goers had gathered in the distance, aware that something might be happening in the llama's yard but not able to see quite what it was or to get any closer.

At last the llamas rose to their feet and Ñusta came over to me for reassurance. The male retreated, looking exhausted.

I managed a quiet word in Ann's ear.

'Was the male successful?'

The animals were so shaggy that it had been difficult to see quite what was happening. I hadn't liked to squat down and peer in the presence of the Curator of Mammals as the children had been doing.

'Yes,' said Ann.

She knew what was what and now I was satisfied that all was well.

Now we had to decide what was the best step to take next. So far the procedure had followed textbook lines. Nothing had happened when we put Ñusta with the male which was constantly with his females but mating had occurred in less than four minutes with the male which was segregated. The scientific papers had said that there was no advantage in repeated matings. If the male were fertile and if the female responded by ovulating, once was enough. Mr Wait was quite happy for us to leave Ñusta at the zoo for a further period but we felt that there might be no need to do this. An animal which did not have regular oestrus cycles might not benefit by being left with the male for another three weeks or a month. Ovulation did not take place until about twenty-six hours after the mating stimulus and, if we took Ñusta home now, she would be safely back in her familiar surroundings before this crucial moment was reached. Also she would not have the stress of a journey in the early stages of pregnancy—enough, we surmised, to trigger off the reabsorption of the embryo which so often happened with llamas. We thought that Ñusta might be unsettled and unhappy away from the comforts of her own home and this might work against the delicate mechanisms of breeding. All had gone very well up to this point, but who could guess what unfortunate mischance might happen if the Um was no longer in our care? We decided that we would take her home with us.

Mr Wait was agreeable to the plan. We arranged to take the children on a quick tour of the zoo and to meet him again at the llama house at half-past four. John's great desire now was to see the bird-eating spider so we hurried off to the Tropical House for a quick view. On that occasion we had no map of the zoo and, among the throngs of people, it was easy to get lost. In the end we managed to see the spider and other creatures briefly as we passed, also to buy some refreshments at the restaurant, but it was nearer to five than half-past four when we returned to the llama house again.

The Curator of Mammals was waiting for us when we arrived. He said that the llamas had mated again in our absence. Ñusta was quite willing to be led out, but she left Wall-eye galumphing round his compound, umming in frustration at the departure of his mate. We bundled Ñusta into the trailer and then Mr Wait on his bicycle led the way to the side gate by which we had entered. We waved goodbye and thanks and then were on the road again.

We had no misadventures on the way home. This time the Land Rover went well but it was dark before we reached Tŷ Mawr. Again Ñusta travelled calmly. It was only later that evening that I noticed the usual slight creases on her cheeks were much more pronounced. It seemed that the day's activities had given her bags under the eyes. The next day she was fine. The bags had gone but she moved a little stiffly. Whether this was the result of the journey or the energetic behaviour of the male, we did not know. She seemed a little detached in manner. Paul said she was due to ovulate at six o'clock that evening but the time came and went and she told us nothing.

Now there was not much that we could do but wait and hope. We knew from our reading that one llama in five failed to ovulate after the mating stimulus. We knew that fifty per cent of those animals which did ovulate would reabsorb the embryo within the first thirty days. Then there were the hazards of the next ten months to face. It was clear that our chances of having an Umlette were a good deal less than fifty-fifty. But of course we hoped.

It was a tantalizing situation. Eleven months seemed a very long time to wait and we wanted to know if there was anything to wait for. I asked Miss Fargher, our vet, if it would be possible to test for pregnancy. She told us that it was possible to make tests but that the practical difficulty was that there was no previous information to which the tests could be related. Llamas in the British Isles were mostly in zoos, usually running with a male. If the animals bred, that was fine but, if not, the zoo would hope for better luck in the following year. Nobody bothered to test llamas for pregnancy in this country. They were not like the more commercially important cattle and horses for which there were well-tried methods. The scientific papers threw little light on this subject. One of them said that alpacas could easily be palpated after four weeks—but could they, we wondered? One felt that animals and their young might be regarded as expendable in the context of research and that one solitary treasure of an Um presented a different problem. We didn't want to know the result of the mating so much that we were prepared to subject Ñusta to unpleasant procedures to find out. There was a practical reason against this too; stress might cause the reabsorption or abortion of the foetus. It would be sad if our curiosity were to destroy the object of our hopes. Mr Ingman, one of Miss Fargher's colleagues, had been making inquiries and had heard that camels—and we supposed this would include the South American camelids too—were more wild than domestic in their breeding habits. Although they had been domesticated for centuries, they still reacted to stress by failing to produce young.

We reviewed the possible methods of pregnancy testing. A rectal examination to detect changes in the uterus and ovaries in the early weeks, and later to palpate the foetus, was possible. This, I felt, would be an enormous shock to Ñusta's dignity, and indeed she might be too small for such an examination, so we ruled it out. Then there was blood-sampling which was frequently used for testing mares. Again we felt that this would upset the llama. Although the collection of blood from the jugular vein was a relatively simple procedure, a series of samples would be needed to detect the changes in hormone levels. The difficulty here would be to know what was 'normal' for a llama. The hormone levels in the blood of a non-pregnant llama would have to be measured as a control.

The collection of urine for tests would be no problem. When I saw Ñusta making for one of her middens it would be a simple matter for me to pop out with something like a saucepan and catch what was needed. She might give me a funny look but there would be no stress. However, it was not known what hormone increases, if any, might be found in llama urine and again we would need a non-pregnant animal as a control.

A useful guide to pregnancy in farm animals is the absence of oestrus behaviour but in the case of llamas this wouldn't help; they didn't have heat periods anyway. Later, in mid-pregnancy, it would be possible to detect the foetal heartbeat with an ultrasonic foetometer which, I was told worked on the depth-sounding principle. Then there was X-radiography, but I felt this advanced technology was not appropriate in our case. Maybe it would be better to wait longer and to rely on the old guides which farmers had used since farming began—weight gain, change of shape and udder development. Many times in the past I had been able to bounce an unborn calf against its mother's body wall during the last two months of pregnancy and feel a hard knobble floating back through the foetal fluids to knock against my fist. This technique, I now learnt, was called 'ballottment' but it was one that I thought Ñusta would not permit. It was sometimes possible to feel the flutter of a foal if you placed a hand just in front of the udder of a pregnant mare after she had had a long cold drink of water. Whether this would work with llamas, I did not know.

Research into methods of pregnancy testing was extremely interesting, but we realized that curious humans could sometimes do more harm than good. Perhaps we should be content to let Ñusta keep her secret. Perhaps nature should be allowed to take its course. In due time the llama would reveal to us whether the journey to Chester had been fruitful. There is a time for all things, and what was a winter and a spring to us when we had seen the leaves in our woods fall and grow again so many times? We would wait as we had done so often in the past.

But the future looked bright. The coming of the llama had, in a way, heralded a new era in the story of Carneddi. For years we had been held back by illness, by the unpredictable state of agriculture and by a galloping inflation which we were ill-prepared to meet. The bringing-up of small children had diverted my time and energy from the farm and, though I loved every minute of being their mother, it was the farm that must support us all. Now Ann and John were older and able to help us with some of our work. The past ten years had stripped us of some of our progress but now it was time to go forward again.

Fred was far away, but I am sure that we were the better for her continuing concern for us. Betty, our first Serf, who had put so much into the farm, came to see us each year though now she was leading her own life. Nothing could compensate for the absence of my father and Mary but they left us the valued legacy of memory. My mother staunchly and ably carried on the family traditions. Beenie was added to our strength and gave all her talents to the life we loved. Now we had Ann and John, children of the mountains, who could ride and run and swim in open spaces, who could eat food grown from our own soil and enjoy a childhood which was not too much debased by technology. This is what Paul and I had worked for. This is what we enjoyed. This was the hard-working, austere but soul-rewarding life of a hill farm. And then, like a catalyst or a touch of magic, the llama came along.